Risk Assessment
The Human Dimension

Risk Assessment
The Human Dimension

Nick W. Hurst
Health and Safety Laboratory, Sheffield, UK

ISBN 0-85404-554-6

© The Royal Society of Chemistry 1998

All rights reserved.

Apart from any fair dealing for the purposes of research or private study, or criticism or review as permitted under the terms of the UK Copyright, Designs and Patent Act, 1988, this publication may not be reproduced, stored or transmitted, in any form or by any means, without the prior permission in writing of The Royal Society of Chemistry, or in the case of reprographic reproduction only in accordance with the terms of the licences issued by the Copyright Licensing Agency in the UK, or in accordance with the terms of the licenses issed by the appropriate Reproduction Rights Organization outside the UK. Enquiries concerning reproduction outside the terms stated here should be sent to The Royal Society of Chemistry at the address printed on this page.

Published by The Royal Society of Chemistry,
Thomas Graham House, Science Park, Milton Road, Cambridge CB4 4WF, UK

Typeset by Land & Unwin (Data Sciences) Ltd, Bugbrooke, Northants NN7 3PA
Printed by Redwood Books Ltd., Trowbridge, Wiltshire

Contents

Author's Notes	vii
Preface	viii
Glossary	xii
An Overview of the Book	xiv

1 Different Perspectives on Accident Causation: Some Accident Case Studies — 1

   1.1 Hardware (Case Studies 1–7) — 2
   1.2 People (Case Studies 8–9) — 7
   1.3 Systems and Cultures (Case Studies 10–12) — 9
   1.4 Summary and Discussion — 11

2 Models of Accident Causation and Theoretical Approaches — 14

   2.1 Some Theoretical Approaches — 14
   2.2 An Analysis of Different Approaches — 32
   2.3 Summary and Discussion — 37

3 The Assessment of Risk – Quantification — 42

   3.1 Engineering Approaches to Risk Assessment — 43
   3.2 Human Reliability Assessment — 48
   3.3 Safety Management Standards and Quantified Risk Assessment — 51
   3.4 Safety Culture and Quantified Risk Assessment — 56
   3.5 Summary and Discussion — 57

4 Risk and Decision Making — 65

   4.1 Risk-based Decisions — 66
   4.2 Measuring Risk Performance Between Sites – Issues of Completeness — 73
   4.3 Summary and Discussion — 77

5 Discussion and Conclusions – Where Does All This Leave Risk Assessment?   80

    5.1 Conclusions from the Previous Chapters   80
    5.2 Risk Assessment – The Human Dimension   83
    5.3 The Subjective/Objective Debate   84
    5.4 Implications for Risk Estimation   87
    5.5 Implications for Risk Evaluation   91
    5.6 Final Discussion   94

Subject Index   100

Author's Notes

1 This book is designed to be read from front to back. It is not a reference book and is designed to be more like a novel; start at the beginning and read to the end. This includes the Preface and Glossary which set the scene for the book. Consequently, the chapters are not self-contained but build one on the other.

2 Accidents are terrible things and no one thinks that an injury or an explosion is about to happen. This book has some case studies in it. I was not connected with any of these and I use them only to illustrate certain points. I am not attributing blame to anyone, and I do not wish people who were involved to be hurt any more by the discussion. But these incidents have occurred and I want to make use of that fact in a manner which I hope will be helpful.

3 I have been fortunate to write this book with funding from the Health and Safety Laboratory's Investment Budget. The Health and Safety Laboratory is an agency of the UK Health and Safety Executive and I am employed by HSE. But this book is a **personal account** of my views of risk assessment. It is not published by HSE but by The Royal Society of Chemistry, of which I am a Fellow, to emphasise its professional nature rather than its official nature.

4 Of course, I am indebted to many people who have helped me with this book. I have had many interesting discussions with friends and colleagues about its contents. I have found these both stimulating and rewarding. I would, in particular, like to mention three people, Pat Twigg, Martin Anderson and John Birtwistle.

Preface

In writing this book I am very aware that we are all engulfed in paper! Every day there is more to read and it is a real problem. I do believe that the average safety professional, student, scientist, engineer or whoever, is expected to know about a range of subjects which is wider than ever and to absorb an ever increasing number of written words. So in writing yet another book what am I trying to achieve? Well brevity (of course) and clarity. I hope this book will be short and clear. To achieve this I have used a bullet point style where possible and lots of diagrams. Some people like pictures and some people don't, but I am in the school of thought which believes that a picture is worth a thousand words.

Also I hope to have something to say which is interesting. My starting point is risk assessment. By this I mean quantified risk assessment as used in the chemical and process industries. The traditions of risk assessment are to do with engineering reliability, statistics, failure rates, consequence models of gas dispersion, pool fires and fireballs. All this sounds a lot like science so it must be right. If it's not right then clearly a bigger computer is needed or some more experiments need to be carried out to ensure the models are okay.

But what about human error? That's important too. People make mistakes. They forget things, misunderstand things and sometimes do something unexpected. Risk assessment can take these human errors into account. What operators need to do is broken down into component steps, and error probabilities assigned to each step. These human error probabilities can be factored into the risk assessment.

Over the past 10 or 15 years, however, the complexity of risk assessment has grown and grown. New ideas have come forward or old ones have been applied in new ways or argued with new conviction. Let me mention some of these.

Safety culture and attitudes to safety have been given a lot of consideration. The view is widely held that a good safety culture is essential to the safe running of complex plant and that a good safety culture is reflected in positive employee attitudes to health and safety issues and strong safety management. These in turn lead to low accident rates.

Safety management systems and the failure of such systems have emphasised the importance of management control in ensuring that proper health and safety standards are maintained. The emphasis in accident causation has been shifted away from individuals doing jobs towards the organisations within which they work and the systems within which they operate. This is reflected in the concepts of the sociotechnical system and systems failures, which have looked widely at why things go wrong. These holistic, or systemic, approaches have considered the allocation of resources, decision making and communication problems as potential causes of accidents in the workplace.

The concepts behind latent failures and active failures have been widely accepted, and contribute to our understanding of accident causation. These ideas, developed by Professor Reason, consider the faults such as poor design which lie hidden (or latent) in an organisation and which, in conjunction with the active or immediate causes of an incident, give rise to failures.

Finally, there are the ideas put forward by the 'high reliability theorists' who believe that serious accidents with hazardous technologies can be prevented through intelligent organisational design and management. This position is often contrasted with the 'normal accident' theorists who believe that accidents are normal and inevitable in complex organisations.

The inspiration for this book comes from the question, "Where does all this leave quantified risk assessment?" If accident causation is so multifaceted, and if causation can go so far back, how is it possible to model such processes at all, let alone in a quantitative manner?

And things get worse! There are important issues concerning the perception of risk, and what constitutes acceptable (or tolerable) levels of risk. It is argued that the best way of viewing acceptable risk is as a decision-making problem and that acceptable risk must be considered in terms of what level of risk is acceptable to whom, when and in what circumstances. In addition risk perception is, at least in part, dependent on personal and group-related variables. These group-related variables are discussed within the social, economic, political and cultural environment within which peoples or individuals live and think. One key idea is that risk is a culturally defined concept. Thus it is widely recognised that the assessment of risk necessarily depends on human judgements and the extent to which risk is a subjective or an objective and measurable 'quantity' is not clear cut. Indeed, there is a wider debate over the contention that social, cultural and political conditions influence and/or determine knowledge and ideas about what is truth i.e. that truth, be it cultural or scientific, need not be objective.

Figure 1 illustrates the above discussion and also defines a 'map' for the book. In the middle of the map is accident causation and risk assessment. There are three main flows into the centre. At the top are technical hardware failures and reliability engineering (Hardware). On the left side are human failures as direct causes of accidents (People) and on the right failures of safety management which arise from the human causes that are

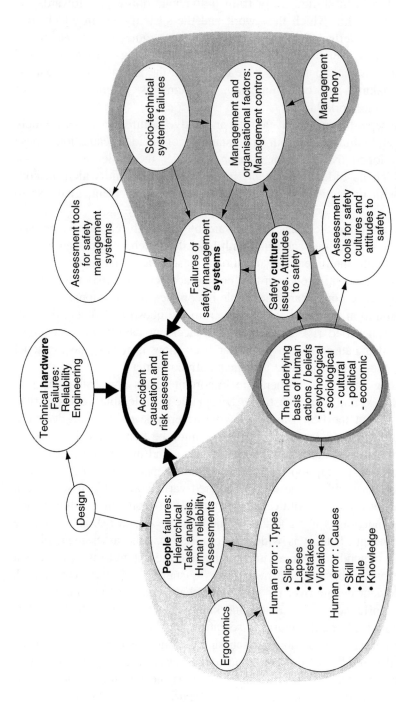

Figure 1 A map of the Book: Risk Assessment - the Human Dimension. In the middle of the map is accident causation and risk assessment. There are three main flows into the centre. At the top are technical hardware failures and reliability engineering (Hardware). On the left side are human failures as direct causes of accidents (People) and on the right failures of safety management systems and cultures which arise from the human causes which are underlying rather than direct (Systems and Cultures)

underlying rather than direct (Systems and Cultures). Figure 1 has some other features. It attempts by use of arrows to show how the various areas are interrelated. An arrow can be 'read' as meaning 'has an effect on' or 'influences' or 'underpins' or 'is related to'. Thus safety culture has an effect on safety management. One aim of the book is to explore the nature of these relationships. Another feature of Figure 1 is that it is wrong! Some arrows are missing. For example, sociotechnical systems influence human failures, and no doubt other arrows are missing. Furthermore, the influences are shown as one way: I have not complicated the figure with double-headed arrows denoting the many reciprocal relationships and feedbacks that in reality flow between the areas represented. But the purpose is to show a model which without being overly complex does shed some light.

The areas related to human failures as a direct cause and human failures as an underlying cause (failures of safety management) are shown as overlapping and are linked via an understanding of the fundamental basis for human actions and beliefs. The figure also shows that many other academic disciplines are relevant here – not just psychology and sociology but ergonomics, reliability engineering, management and organisational theory and systems theory. To say that health and safety is a multi-disciplined science would seem to be an understatement, because many of these disciplines are multi-disciplined themselves and are underpinned by mathematics, physics and chemistry for example.

So Figure 1 is a map of the book which is presented as five chapters. Each chapter will try to further develop either the whole map or part of it. The map of the book is used as a symbol throughout the book to tell you exactly what is the topic area under consideration.

Glossary

The concepts behind the following definitions and the development of these concepts are explored in this book. These definitions, therefore, are *a starting point* for the book, not its end product.

The terminology associated with the terms 'hazard' and 'risk' are a particular source of difficulty. A hazard is defined as an object or a situation with the potential to cause harm. Thus a hazard in the workplace might be a chemical or a machine. Anything which can hurt people if certain circumstances prevail, e.g. a chemical is spilt and inhaled, or contact is made between a person and a moving machine, is a hazard. The concept of risk tries to go beyond hazard and to answer the following questions: (1) how likely is something to go wrong? and (2) what will the effect be? So the definition of risk contains the two elements, frequency (or probability) and consequence.

Risk tries to describe both the likelihood of an event and the consequence. But in order to compare risks it is necessary to 'fix' the consequence. For example, if the consequence considered is 'death to an individual' then the risks from a series of activities can be compared by considering the frequency with which these activities cause death. Similarly, if the consequence chosen is an injury resulting in (say) three days lost from work, then risks for a series of activities can be compared by considering the frequency with which the activities cause a 'three-days-lost' accident. However, if one activity is likely to cause death (e.g. by entrapment in a machine) but the likelihood is very low and another activity is likely to cause a less serious injury (e.g. a three-days-lost accident) but with a higher frequency then no simple and unambiguous method is available to compare the two risks.

Acceptable Risk The concept of whether a risk is acceptable, or not, to a group or an individual and in what circumstances the risk is considered acceptable.

Hazard The situation or object that in particular circumstances can lead to harm, i.e. has the potential to cause harm.

Human Reliability Assessment (HRA) The process of estimating human error probabilities for use in risk assessment.

Glossary

Post Normal Science Human activities in which scientific evidence and experience are important but the facts are uncertain, values in dispute, stakes high and decisions urgent.

Probabilistic Risk Assessment (PRA) or Probabilistic Safety Assessment (PSA) Terms often used in the nuclear industry for the process of making quantified estimates of risk and their evaluation.

Quantified Risk Assessment (QRA) A term used for the quantified estimation and evaluation of risk, often in the chemical industry.

Risk The probability that a particular adverse event occurs during a stated period of time, e.g. risk of death while hang-gliding during a seven-day period. This would be small for a randomly selected inhabitant of the UK, but its value will alter according to age, season, weather and membership of a hang-gliding club.

Risk Assessment The study of decisions subject to uncertain consequences. It consists of Risk Estimation and Risk Evaluation.

Risk Estimation How big is the risk?

Risk Evaluation How significant is the risk and hazard and to whom?

Risk-based Decision Making A process in which a decision is made to proceed, or not to proceed, with a course of action in which the level of risk is relevant to the decision which is taken (see Acceptable Risk).

Risk Management The making of decisions concerning risk and the subsequent implementation of the decisions which flows from risk estimation and risk evaluation.

Risk Perception The perception that an individual or group has about the extent of a particular risk. Is the risk perceived to be great or small.

Safety Relates to the freedom from unacceptable risks of harm to people or a person either locally to the hazard, nationally or even world wide.

Safety Culture The attitudes, beliefs and feelings people in an organisation have towards health and safety.

Safety Management The application of organisational and management principles to achieve optimum safety with high confidence, i.e. management applied to achieving safety.

Social Amplification of Risk How estimates of risk are enhanced or reduced by social processes, i.e. how the risk is perceived. An individual, or a society, may attach particular significance to a particular risk.

Socio-political Amplification of Risk How social and political processes enhance actual risk to individuals in particular circumstances, e.g. in industrialising countries.

An Overview Of The Book

CHAPTER 1 DIFFERENT PERSPECTIVES ON ACCIDENT CAUSATION: SOME ACCIDENT CASE STUDIES

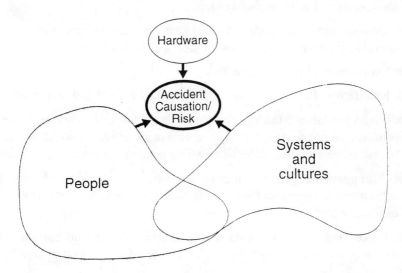

Figure 0.1 *The three main contributors to accident causation and risk assessment. Different perspectives are illustrated: hardware failures, failures of people and failures of systems and cultures*

When an accident or incident occurs which is of sufficient scale to attract public concern, for example a train accident or chemical incident, the event is viewed from different perspectives by different people and various 'causes' are attributed. The purpose of this chapter is to explore these different perspectives. Some people will view the accident as 'human error', others as 'management failures' or failures of the 'safety culture' and others from a technical viewpoint. These positions are examined and set against each other in this chapter. Is one perspective more valid than another? A case study approach is used to illustrate the different viewpoints.

CHAPTER 2 MODELS OF ACCIDENT CAUSATION AND THEORETICAL APPROACHES

Because of the different perspectives explored in Chapter 1 a range of models and theoretical approaches exist to 'explain risk' and the causes of accidents. Some of these are very specific and consider, for example, different types of human error, while others attempt to be holistic or systemic in their approaches. One main purpose of this chapter is briefly to describe this range of theoretical work and accident causation models.

Failure of Hardware

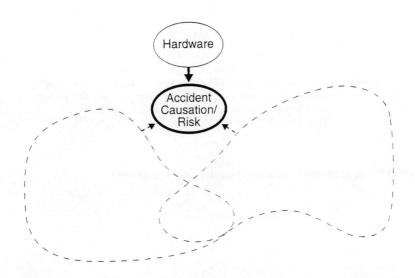

Figure 0.2 *Some models of accident causation emphasise failures of hardware, while failures of people and failures of cultures and systems are not emphasised*

Direct Human Failures (People)

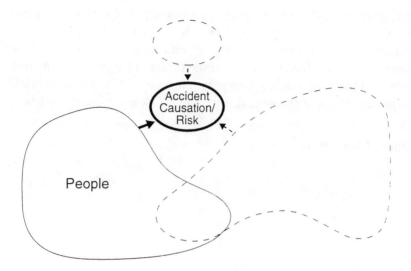

Figure 0.3 *Some models of accident causation emphasise the failures of people as direct causes of accidents, while the failures of hardware or systems and cultures are not emphasised*

Underlying Human Causes (Systems and Cultures)

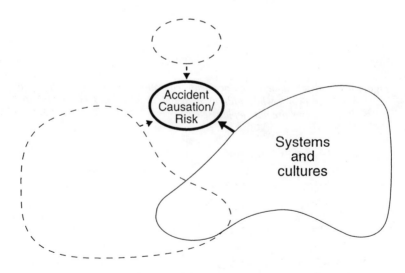

Figure 0.4 *Some models of accident causation emphasise the failures of systems and cultures as underlying causes of accidents, while the failures of hardware or people are not emphasised*

- Safety cultures, attitudes to safety
- Failures of safety management systems
- Sociotechnical systems failure
- Latent failures and active failures
- High reliability theory/normal accidents

The different approaches are cross referenced to the case studies in Chapter 1 and a simple attempt is made to draw together the different approaches, looking for overlap and commonalities. It is shown that a range of organisational theories describes 'ideal' organisations while another group of theories describes 'real world' organisations.

A second main purpose of the chapter is to introduce the complexity of risk under the broad heading of an objective/subjective debate in risk assessment and to consider issues of risk perception and acceptability.

- The subjective/objective debate
- Risk perception/acceptable risk

CHAPTER 3 THE ASSESSMENT OF RISK – QUANTIFICATION

When risk is analysed and quantified there is an underlying assumption that the methodology used for the risk assessment is 'correct' and complete, i.e. that it represents the system it purports to model. Thus a quantified risk assessment of a chemical plant involving hardware failures, modelling of fires and explosions, etc. is taken to be a good model of the system and hence to generate realistic values of the risk. It is assumed that appropriate data are available to use within the models. The first two chapters of the book have laid the ground that a risk assessment which considers only hardware failures and ignores safety management, safety culture and human error is not to be considered complete. Because of this, it is important to establish links between safety cultures, human errors, safety management and risk assessment, and this chapter describes in general terms the efforts which have been made to develop tools which make these connections explicit.

Engineering Approaches to Risk Assessment

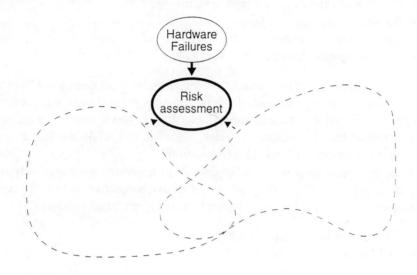

Figure 0.5 *Some models for risk assessment emphasise the failures of hardware, while the failures of people, systems and cultures are not emphasised*

Human Reliability Assessment

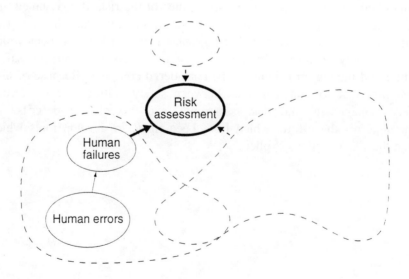

Figure 0.6 *Methods for human reliability assessments have been developed. These methods provide a quantified estimate of human error probabilities which can be factored into a risk assessment*

Safety Management Standards and Quantified Risk Assessment

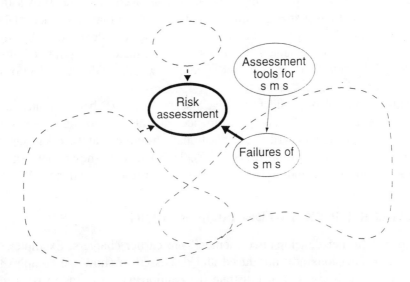

Figure 0.7 *Assessment tools for safety management systems (SMS) have been developed. These tools provide a measure of the quality of the safety management system which in turn affects the risk*

Safety Culture and Quantified Risk Assessment

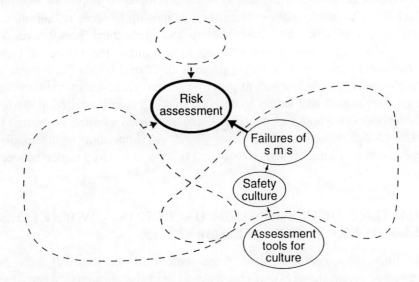

Figure 0.8 *Assessment tools for safety culture have been developed. These tools provide a measure of the strength of the safety culture which in turn affects the risk*

This chapter further develops the subjective/objective debate introduced in Chapter 2, discusses the complexity of risk as a concept and shows that risk assessment has a strong human dimension which may make acceptable risk decisions difficult to arrive at. Risk is neither objective nor subjective absolutely and the chapter describes risk estimates in terms of their 'relative objectivity' to indicate the extent to which evidence and information enter into the risk estimate.

This chapter also considers relative risk performance between sites, and introduces a scale of risk performance in relation to the theoretical perspectives which describe organisational structure, safety management and safety culture (Sections 2.1.3–2.1.7). The risk scale shows low risk for 'ideal' organisations and higher risk for 'real world' organisations.

CHAPTER 4 RISK AND DECISION MAKING

Chapter 4 further develops two themes from earlier chapters. Examples of risk-based decisions are introduced and discussed to show the complexity of risk as a concept and to illustrate the caution which needs to be taken when risk estimates are used in decision making. Secondly, the overlap and commonalities which have been shown to exist between different approaches and descriptions of 'ideal' and 'real world' organisations are further developed by considering how, when the differences in 'idealness' between two sites are measured, the measurement might be related to risk performance, i.e. it discusses the relationship between 'idealness' and risk. The chapter refers to the auditing systems developed by European research, which have helped to make explicit the relationship between organisational 'idealness' and risk. The case studies and theoretical perspectives of Chapters 1 and 2 are also drawn upon to explain the nature of these systems and their application. The issue is discussed within the context of completeness in risk assessment processes. Finally, the chapter shows that in using estimated risk values in decision making, care is needed in the use of poorly defined and understood risk concepts such as those embodied in risk ranking approaches. For example, it is important to distinguish between risk ranking on a single site and relative risk performance between sites.

CHAPTER 5 DISCUSSION AND CONCLUSIONS – WHERE DOES ALL THIS LEAVE RISK ASSESSMENT?

The final chapter draws together the themes of the book. It uses the conclusions from the previous chapters and asks the question, "where does all this leave risk assessment?" The conclusions are discussed in terms of the subjective/objective debate in risk assessment and the human dimensions of risk assessment. The implications of the conclusions for the practice of risk estimation and the evaluation of risk are also both considered. The chapter concludes with an agenda for risk assessment over the next few years. This agenda attempts to describe important areas for the development of risk assessment based firmly on the conclusions of this book.

CHAPTER 1

Different Perspectives on Accident Causation: Some Accident Case Studies

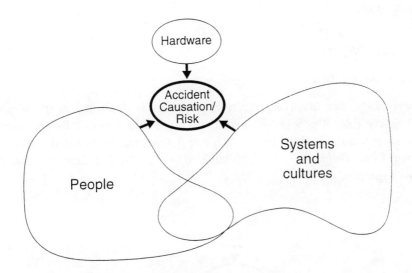

Figure 1.1 *The three main contributors to accident causation and risk assessment. Different perspectives are illustrated: hardware failures, failures of people and failures of systems and cultures*

Table 1 lists some well known case studies. These are events or accidents which have been in the public eye at some time and have been the subject of investigations, enquiries, newspaper reports and so on. They are drawn from various areas: aircrashes, failures of chemical plant, a football stadium disaster and others. Generally, there are detailed published accounts and analyses of these events which describe how they happened. These are listed in the references. Table 1 also lists some different perspectives that have been taken, at various times, by others in considering the causes of the events. These are illustrated in Figure 1.1. My purpose here is to look briefly at these perspectives for the events described. At this stage I am not trying to show that some may be right and others wrong, but simply to show that different individuals and organisations tend to take different views. This raises the question as to whether a search for more complete descriptions might not suggest practical consequences for risks and safety.

Although the case studies are drawn from various areas, I am assuming that the lessons which can be learnt are independent of the area and can be applied to other areas, i.e. I assume that the underlying causes of accidents, events and incidents will cover similar, general types in the various areas considered. Consequently I have not chosen chemical plant incidents, exclusively, but events or incidents which I consider best illustrate the different perspectives. I assume that the conclusions drawn will be applicable to chemical plant incidents in a general sense, e.g. that weak management will lead to problems irrespective of the area which is managed weakly.

1.1 HARDWARE

1.1.1 Case Study 1

A warehouse fire involving reactive chemicals is described in the Health and Safety Executive's (HSE) investigation into the fire at Allied Colloids Ltd, Bradford on 21 July 1992.[1] In Table 1 this incident is listed under the perspectives *Technical issues highlighted – How did it happen?* and *In what sequence?*

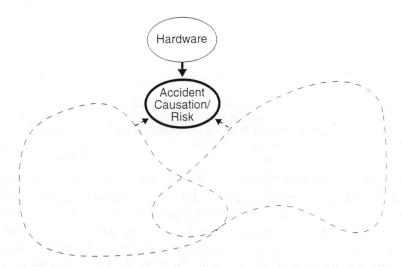

Figure 1.2 *Accident causation in which the role of hardware failures is emphasised. This might be a direct cause, e.g. poor welding, or an underlying cause, e.g. poor design. Sequences of events may be investigated or different explanations tested and proved or disproved*

Perspectives on Accident Causation: Some Case Studies

Table 1 The accident case studies listed have been considered at various times from different perspectives. The symbol ● is used to show where a particular perspective has been noted

Case Studies	Perspectives						Technical issues highlighted		
	Failures of management	Complete investigations	Human error	Failure of safety culture	How did it happen?	What sequence?	Hardware failures Direct	Hardware failures Underlying	Alternative explanations
1 Allied Colloids[1]					●	●			
2 Ammonia/Air Mixer[2]					●	●			●
3 Hillsborough[3]					●		●		●
4 Spanish Campsite[4,5]									
5 Ammonia Lithuania[6]									
6 Feyzin[7]								●	
7 Potchefstroom[8]							●		
8 Clapham Junction[9,10]		●	●						
9 Kegworth Air Crash[11–16]		●	●						
10 Escape from Parkhurst[17,18]	●								
11 The Herald of Free Enterprise[19]	●								
12 Wylfa Nuclear Power Station[20,21]				●					

The HSE investigation report includes lessons learnt and considers the management of health and safety, but emphasises both the possible sequences of events and the scientific basis used to explain how the fire happened. The following points from the report give an indication of the nature of the investigation.

- Numerous employees who had been directly involved with the incident were identified and interviewed.
- Efforts were concentrated on gaining an appreciation of the range and quantities of chemicals involved and how and where they were stored.
- The scientific investigation examined the properties of and interactions between the materials stored, the nature of the products of combustion and the general spread of the fire. Extensive studies were carried out on a number of chemical samples from the site.
- The investigation also examined potential sources of heat capable of raising the temperature of the contents of reactive chemical kegs to the point where sufficient material would decompose to cause that package to fail.
- The investigation looked at the failure characteristics of the kegs and the spread of the fire through the warehouse. A technique known as computational fluid dynamics was used to help quantify the heat sources.

HSE investigations soon established that the stores contained a self-reactive substance which was thermally unstable and capable of undergoing violent decomposition at relatively low temperatures. The incident started when two or three kegs of this substance ruptured. These were stored on the top shelf of the racking in the warehouse, close to the steam condensate return line and a roof light panel. The sun would not have been shining directly on the kegs, and it was concluded that a malfunction of the steam heating system or operator error caused the condensate pipe to be hot.

1.1.2 Case Study 2

Other incident investigations, such as the catastrophic failure of an ammonia/air mixer described by Verduijn follow a similar form.[2] There is an emphasis on establishing the sequence of events and the direct technical causes, which include the chemical reactions involved and any failures of chemical plant items and also on describing lessons learnt.

1.1.3 Case Study 3

The investigation of the Hillsborough Football stadium disaster by the Health and Safety Executive[3] focused on technical aspects of the disaster,

Perspectives on Accident Causation: Some Case Studies 5

including metallurgical examination of crash barriers, collapse load calculations and development of a model to predict crowd pressures, among other issues. Ninety-five people died from crush asphyxia because of severe overcrowding. The investigation provided important evidence in determining what happened at Hillsborough and enabled the elimination of some of the theories put forward in the aftermath of the disaster.

1.1.4 Case Study 4

Clearly, after a tragic incident there is a desire to find out what happened. This can sometimes lead to alternative technical explanations of the cause of the incident (even after the court of enquiry into the event). On 11 July 1978 a disaster occurred at a campsite in Spain in which over 200 people lost their lives due to fire and explosions involving a road tanker carrying liquefied petroleum gas which burst open and lost its contents. The exact events which occurred were the subject of a debate in the literature[4,5] which centred around whether or not the tanker exploded due to overloading followed by sunshine raising the tank temperature; whether a vehicle crash occurred; or, finally, whether the tanker was first engulfed in a fire which subsequently caused the explosion. Such debates and investigations are important because the outcome can affect subsequent design and operation of hazardous technologies.

1.1.5 Case Study 5

Alternative explanations also arose in the case of an ammonia tank failure in Lithuania.[6] On 20 March 1989, a large storage tank containing refrigerated liquid ammonia failed suddenly without any warning and moved sideways, demolishing a reinforced concrete bund wall and releasing 7000 tonnes of liquid ammonia. The ammonia vapour ignited and the resultant fire spread to involve 35,000 tonnes of compound fertiliser in nearby warehouses. Acid fumes from the burning solid continued for three or four days and the plume was seen from some 45 km distance, the clouds being discoloured with nitrous fumes many kilometres downwind. There were seven fatalities reported at the time of the incident and 57 reported injuries on site, most being gassed by the ammonia which formed a pool of liquid up to 70 cm in depth over a wide area. There were no reported fatalities outside the plant. Subsequent debate centred around the question of whether the event had occurred due to 'rollover', in which warm liquid at the bottom of the tank rises while cold liquid settles, the process causing the rupture of the tank, or whether the tank failed due to overpressure because of poor vent design.

1.1.6 Case Study 6

As mentioned above, it is important to establish the technical and operational detail of accident causation when the lessons learnt affect the subsequent design and operation of hazardous plant. The propane fire at Feyzin in France in 1966 is a case in point.[7] A cloud of propane vapour spread 160 m until it was ignited by a motor car on the adjoining motorway. The liquid spillage burnt underneath the tank and over-heated it until it burst. A wave of burning propane then engulfed the firemen while debris broke the legs of the next tank and spread the fire. Major weaknesses in the design of such installations were not appreciated until after the Feyzin disaster. In addition, the method of operation to be used with the improved design could be clearly specified. The Feyzin fire is listed in Table 1 under the perspectives of underlying hardware failures, where the cause of the accident is considered to be inadequate design of the plant, i.e. a hardware failure which is underlying rather than direct.

1.1.7 Case Study 7

Case study 6 can be contrasted with the ammonia tank failure at Potchefstroom, South Africa in 1973 when a 50 tonne storage tank for liquid ammonia failed.[8] The fertilizer plant had four 50 tonne capacity horizontal bullet type ammonia pressure storage tanks. One of these tanks failed without warning, with a section approximately 25% of the cross sectional area coming out from one dished end of the 2.9 m diameter × 14.3 m long vessel as a result of brittle fracture. An estimated 30 tonnes escaped from the tank plus another 8 tonnes from a tank car. The gas cloud that was immediately formed was about 150 m in diameter and nearly 20 m in depth. The ammonia cloud caused the deaths of eighteen people including six outside the plant fence. Approximately 65 people required medical treatment in hospital and an unknown number were treated by private doctors. The end of the tank which failed had been repaired by welding and the welded area had not been stress relieved. Therefore, the most likely cause of failure was brittle fracture. In Table 1 this is called a direct hardware failure, as opposed to an underlying design problem.

The case studies on hardware failures therefore emphasise the technical and scientific causes of accidents. The sequences of events are established and plausible explanations are explored and set one against the other. Often samples are removed from the site and tested for their chemical and metallurgical properties. Often important lessons for the future are learnt in terms of design, material specification and scientific mechanisms. Scale models may be constructed and tested so that the best explanations for the incident become apparent.

1.2 PEOPLE

1.2.1 Case Study 8

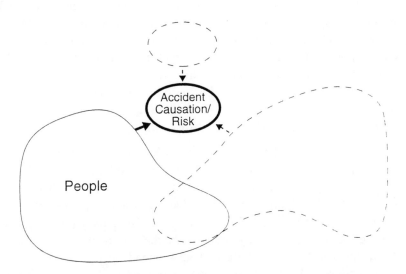

Figure 1.3 *Accident causation in which the failures of people are explored not only as direct causes of accidents but also implicating the reasons for the underlying causes of people's behaviour*

The investigation into the Clapham Junction Railway Accident (12 December 1988, the Hidden report[9]) is listed in Table 1 under the perspectives of 'Complete Investigations' and 'Human Error'. This is because, to quote the report, "The direct cause of the accident was undoubtedly the wiring errors made by a specified individual in his work in the Clapham Junction 'A' relay room (human error is a direct cause) ... a welter of criticism was justifiably laid at the door of this individual". But the report is also very broad in its approach. To quote Sir Bob Reid,[10] "The immediate cause was some faulty wiring work. It was not particularly hard to pinpoint one or two individuals who were at fault. But lying behind the accident and lying behind individual errors was a whole chain of circumstances that has everything to do with management responsibility." The Hidden report covers, for example, training, supervision, working practices, hours of work and even pay and reward structures. At the time of the accident, MPs questioned whether there was a link between the accident and financial cutbacks at British Rail. Thus the incident highlighted human error as a direct cause, but later investigations introduced the concepts of weak safety culture, underlying causes of failure and failures of management.

1.2.2 Case Study 9

Similarly, the Kegworth air disaster on 8 January 1989 implicated human error as a direct cause. It became clear that the engine on the port (No 1 or Left) side of the aircraft had caught fire but the starboard engine (No 2 or Right) had been shut down, in flight, by the pilots rather than having failed. A spokesman for the company which made the engines said that a pilot shutting down an engine would have to carry out two or three separate actions. A shutdown would have to be a series of deliberate acts.[11]

Much later (February 1996) the *Daily Mail* ran a headline "Pilot blamed for jet disaster gets £100,000".[12] The story ran:

> A pilot partly blamed for the Kegworth air disaster which killed 47 people has won more than £100,000 compensation, it emerged yesterday. The Captain was awarded the cash by British Midland's insurers after he claimed the airline had not trained him properly on the twin-engined Boeing 737 he was flying. An enquiry into the crash showed he and his co-pilot shut down the wrong engine moments before the Heathrow to Belfast plane plunged on to a motorway in Leicestershire, UK. The Captain was paralysed in the accident seven years ago and is confined to a wheelchair. His solicitor refused to disclose the exact amount of the award, saying only that it was a 'substantial' six-figure sum. He added: 'We are reasonably satisfied.' The crash inquiry heard that the 737 was preparing for an emergency landing at East Midlands airport after one engine caught fire. But instead of shutting down the left-hand one which was engulfed in flames, The Captain and First Officer stopped the working right-hand engine. The aircraft, with 125 people on board, belly-flopped into a field, then ploughed across the motorway and was embedded in an embankment. The Captain claimed compensation for personal injury on the grounds that the plane developed a fault which led to the crash and he had not been given simulator training on how to deal with the problem. A Transport Department report[13] on the disaster criticised The Captain and First Officer for acting against their training when they shut down the wrong engine. It said: "Their misdiagnosis must be attributed to their too-rapid reaction and not to any failure of the engine instrument system." But the reasons behind the error were never fully established. The Captain told the crash inquest: "I never doubted that we had got the correct engine." And the First Officer was convinced the new aircraft's electronic instrumentation panel wrongly indicated that the right-hand engine was on fire.

The Kegworth air crash is analysed by Denis Smith[14] who says that while the pilots of the aircraft have shouldered the bulk of the blame to date (1992) it can be argued that the roots of the crash are much more complex than simply pilot error. Bill Richardson[15] makes this point well.

There are a range of views on the issue of what causes disasters. These range from 'simple-causal' explanations to 'complex systems' explanations. The 'simple' view holds that 'It is somebody's fault. A simple activity would have avoided this'. This view sees the problem as being essentially one of human failing, i.e. people (behaving negligently) do things or fail to do things and so create disasters.

However, there is an altogether more 'messy' view of disaster causation. In this 'messier' view, complex systems of beliefs, power, economics, social relationships, technologies, management systems and nature, for example, in both the organisation and its environment, interact with one another to create a complex, interactive system which is naturally prone to what Perrow[16] has termed 'normal accidents'.

This complexity is clearly illustrated by the *Daily Mail* article quoted above in which the pilot 'blamed' for the accident receives compensation for his injuries. Thus the Kegworth incident implicated human error as a direct cause of the incident but also clearly shows the complexity of these issues and how underlying causes contribute to failure.

1.3 SYSTEMS AND CULTURES

1.3.1 Case Study 10

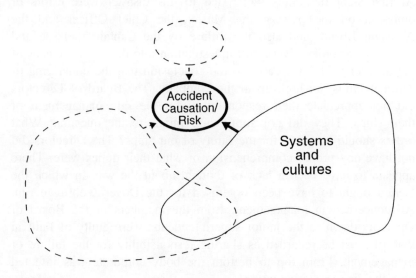

Figure 1.4 *Accident causation in which the failures of systems and cultures is emphasised. These might be safety management systems or safety cultures*

The escape of three prisoners from Parkhurst Prison on the Isle of Wight on 3 January 1995 is analysed by the Learmont report.[17] The report concludes that Parkhurst should never have been retained as a high security prison. The procedures employed were lax and unprofessional, made worse by the disastrous indecision and apathy which pervaded so many working practices. The escape revealed a chapter of errors at every level and naivety that defies belief. The many shortcomings at Parkhurst illustrate the failure of the prison service to learn the lessons of the past. Thus, the escape was not due to one person's folly, because many of its ingredients can be traced along the lines of communication to prison service headquarters. The numerous failings identified in this report indicate that there were many hands on the tiller on this voyage to disaster. *The Guardian*[18] summarised this with the headline "Damning verdict on prison chiefs", which clearly implies a failure of management. In addition the idea of different levels of responsibility and failures of communication between them is clearly shown by the report.

1.3.2 Case Study 11

The Zeebrugge car ferry disaster involving the capsize of the *Herald of Free Enterprise* occurred on 6 March 1987 with the loss of 188 lives. A formal investigation is presented in the Sheen report.[19] The report states that:

> At first sight the faults which led to this disaster were errors of omission on the part of the Master, the Chief Officer and the Assistant Bosun, and also the failure by the Captain to issue and enforce clear orders. But a full investigation into the circumstances of the disaster leads inexorably to the conclusion that the underlying or cardinal faults lay high up in the company. The Board of Directors did not appreciate their responsibility for the safe management of their ships. They did not apply their minds to the question: What orders should be given for the safety of our ships? The Directors did not have any proper comprehension of what their duties were. There appears to have been a lack of thought about the way in which the Herald ought to have been organised for the Dover/Zeebrugge run. All concerned in management, from the members of the Board of Directors down to the junior superintendents, were guilty of fault in that all must be regarded as sharing responsibility for the failure of management. From top to bottom the body corporate was infected with the disease of sloppiness.

In Table 1 case studies 10 and 11 are listed under the perspective 'Failures of Management'. These case studies include the ideas of direct causes and underlying causes of accidents, multiple causality and the concept of 'levels' of causation. This perspective illustrates how failures that are remote in both time and place from the event can nevertheless be seen as 'causes' of the incident. There is a clear understanding that management failures can lead to 'front line' failures. This is illustrated in Figure 1.4 which shows failure of systems and cultures as causes of accidents. These might be failures of safety management systems or safety cultures.

1.3.3 Case Study 12

Table 1 lists the Wylfa nuclear power station incident under the perspective of safety culture failure.[20, 21] A steel grab on the end of a crane used to lift nuclear fuel rods broke off and fell into a reactor because of a faulty weld. In a statement read to the court during proceedings at Mold Crown Court, the Chief Inspector of Nuclear Installations said the event was potentially the most serious in the United Kingdom during his time (as Chief Inspector) and that he was particularly concerned about the blatant failure of the safety culture. The court had heard that officials appeared to put money before safety. The judge in his summing-up said that the incident was the fault of the company and not of individuals, but rejected the idea that action had been deferred for commercial or contractual considerations and that safety was not the prime consideration in the minds of those in charge. It is interesting to note that the court at least in part considered aspects of organisational culture and the possibility of safety being compromised by commercial considerations as underlying causes of the incident. These are the types of consideration which are described by the safety culture of an organisation and illustrated in Figure 1.4, in which failures of culture are shown as causes of accidents.

1.4 SUMMARY AND DISCUSSION

This chapter has examined a number of case studies which are briefly summarised under different perspectives in Table 1. Sometimes human error is implicated as a direct cause of the incident but often the human errors are underlying problems – poor safety management systems or poor safety culture. Hardware failures also occur. These may be direct causes of failure as in structural collapse or brittle fracture, but may also be underlying causes such as poor design, for example. Often official enquiries into important incidents seek to obtain a detailed and complete understanding of all the relevant contributing factors. The pursuit of 'completeness' is considered to be important because it provides a more rich account of the event and so helps to ensure that lessons are learnt for the future.

At other times accounts of incidents seem to emphasise one particular point of view. It is not always clear why this should be the case but the expertise and interests of individuals and of organisations may be relevant. However, such accounts will tend to be incomplete and lack richness. Important lessons may be missed. This is an important human dimension to accident investigation because the direction of the investigation may be determined by the expertise and interests of the organisations and the individuals involved.

In general terms in accident case studies the following issues are likely to be important:

- How did it happen?
- What was the sequence of events?
- What were the direct hardware causes?
- What were the underlying hardware causes?
- What were the direct human causes?
- What were the underlying human causes?

A complete account would seek to look at all these aspects to learn lessons for the future. Case studies 1–7 look at aspects of hardware failure and consider issues such as "How did it happen?" and "What was the sequence of events?" Case studies 8 and 9 look at incidents in which human error as a direct cause is implicated and discussed. Case studies 10–12 describe aspects of underlying human causes of accidents, including a discussion of safety culture issues and failures of management. Thus the case studies illustrate how different perspectives have been taken at various times by various people, in considering the causes of accidents and incidents.

FURTHER READING

1. J. Lancaster, *Engineering Catastrophes – Causes and Effects of Major Accidents*, Abington Publishing, Cambridge, 1996.

2. B. Toft and S. Reynolds, *Learning from Disasters – A Management Approach,* Butterworth-Heinemann, Oxford, 1994.

3. D. Vaughan, *The Challenger Launch Decision: Risky Technology, Culture and Deviance at NASA*, University of Chicago Press, Chicago, 1996.

REFERENCES

1 Health and Safety Executive, *The Fire at Allied Colloids Limited, HSE Report of Incident 21/7/1992*, HMSO, London, 1993.
2 W. D. Verduijn, *Process Safety Progress*, 1996, **15**, 89.
3 C. E. Nicholson and B. Roebuck, *Safety Science*, 1995, **18**, 249.
4 I. Hymes, Loss Prevention Bulletin No. 61, 1985, 11.
5 V. C. Marshall, Loss Prevention Bulletin No. 72, 1986, 9.
6 B. O. Anderson, Loss Prevention Bulletin No. 107, 1989, 11.

7. Anonymous, Loss Prevention Bulletin No. 87, 1987, 1.
8. H. Lonsdale. Ammonia Tank Failure – South Africa Ammonia Plant Safety. AIChE 1974. pp 126-131.
9. Department of Transport, *Investigation into the Clapham Junction Railway Accident* (The Hidden Report), HMSO, London, 1989.
10. B. Reid, *Safety Management*, November 1993, p.12.
11. *The Independent*, London, 11 January 1989.
12. Nick Hopkins, *Daily Mail*, London, 2 August 1996.
13. Air Accident Investigation Branch, *Report on the accident to Boeing 737-400.G-0 Brie near Kegworth Leicestershire on January 8th 1989, Department of Transport Aircraft accident Report 4/90*, HMSO, London, 1990.
14. D. Smith, Disaster Management, 1992, **4**, 63.
15. Bill Richardson, *Disaster Prevention and Management*, 1994, **3**, 41.
16. C. Perrow, *Normal Accidents: Living with High-risk Technologies*, Basic Books, New York, 1994.
17. Sir John Learmont, *The Escape from Parkhurst Prison 3/1/1995* (The Learmont Report), HMSO, London, 1995.
18. Sarah Boseley, *The Guardian*, London, 17 October 1995, p.4.
19. Department of Transport, *The Merchant Shipping Act, 1894, MV* Herald of Free Enterprise, *Report of court No. 8074* (The Sheen Report), HMSO, London, 1987.
20. Daniel John, *The Guardian,* London, 13 September 1995, p.1.
21. Godric Jolliffe, *Safety Management*, October 1995, p.1.

CHAPTER 2

Models of Accident Causation and Theoretical Approaches

Because of the different perspectives explored in Chapter 1 a range of models and theoretical approaches exists to 'explain risk' and the causes of accidents. Some of these are very specific and consider, for example, different aspects of engineering design or different types of human error, while others attempt to be holistic or systemic in their approaches. One main purpose of this chapter is to describe briefly this range of theoretical work and accident causation models. The different approaches are cross referenced to the case studies in Chapter 1 and a simple attempt is made to draw together some of the different approaches looking for overlap and commonalities.

The second purpose of the chapter is to introduce the complexity of risk under the broad heading of an objective/subjective debate in risk assessment, and to consider issues of risk perception and acceptability.

2.1 SOME THEORETICAL APPROACHES

2.1.1 Engineering Considerations

Health, safety and loss prevention are important aspects of the engineering sciences. In describing, for example, the discipline of chemical engineering considerable emphasis is placed on safety and operability of the plant as well as efficiency of the design and process in ensuring economic production.[1]

Health, safety and loss prevention from the engineering perspective can be considered under the following broad headings:

- Identification and assessment of the hazards, e.g. toxic or flammable materials.
- Control of the hazards, e.g. by containment of flammable and toxic materials.
- Control of the process, e.g. prevention of hazardous deviations in process variables (pressure, temperature, flow) by provision of automatic control systems, interlocks, alarms and/or, trips, together with good operating practices and management.

- Limitation of the loss, e.g. minimising the damage and injury caused if an incident occurs by pressure relief, the provision of fire-fighting equipment and careful plant layout.

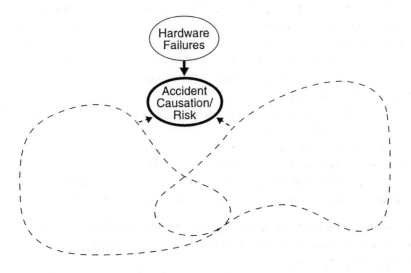

Figure 2.1 *Some models of accident causation emphasise failures of hardware while failures of people and failures of cultures and systems are not emphasised*

Chemical processes can be divided into those that are intrinsically safe, and those for which the safety has to be engineered in. An intrinsically safe process is one in which safe operation is inherent in the nature of the process; a process which causes no danger, or negligible danger, under all foreseeable circumstances. Clearly, the designer should always select a process that is inherently safe whenever it is practical, and economic, to do so. However, most chemical manufacturing processes are, to a greater or lesser extent, inherently unsafe, and dangerous situations can develop if the process conditions deviate from the design values. The safe operation of such processes depends on the design and provision of engineered safety devices, and on good operating practices, to prevent a dangerous situation developing, and to minimise the consequences of any incident that arises from the failure of these safeguards.

The term 'engineered safety' covers the provision in the design of chemical plant of control systems, alarms, trips, pressure-relief devices, automatic shut-down systems and duplication of key equipment and services, and of fire-fighting equipment, sprinkler systems and blast walls, to contain any fire or explosion.

In considering the hazards associated with chemical engineering processes mention can be made of toxicity (short term and long term), flammability and explosive potential. Consequence models exist to describe the likely effects of the realisation of these hazards on people, property and the environment. For example a 'pool fire' model would calculate thermal radiation as a function of distance from a burning pool of (say) aviation fuel. The effects of the thermal radiation on people and property can be considered by the use of separate relationships or tables of damage. A consequence model for a vapour cloud explosion would consider the mass and type of substance involved and the extent to which air is entrained into the fuel, and calculate overpressure and thermal radiation levels as a function of distance from the explosion.

Mention should be made of the hazard evaluation method developed by the Dow Chemical Company. To assess the potential hazards of a new plant the method can be used after the piping and instrumentation and equipment layout diagrams have been prepared.[1]

The basic safety and fire protective measures that should be included in all chemical process designs based on the Dow Guide, include:

- Adequate, and secure, water supplies for fire fighting.
- Correct structural design of vessels, piping, steelwork.
- Pressure-relief devices.
- Corrosion-resistant materials, and/or adequate corrosion allowances.
- Segregation of reactive materials.
- Earthing of electrical equipment.
- Safe location of auxiliary electrical equipment, transformers, switch gear.
- Provision of back-up utility supplies and services.
- Compliance with national codes and standards.
- Fail-safe instrumentation.
- Provision for access of emergency vehicles and the evacuation of personnel.
- Adequate drainage for spills and fire-fighting water.
- Insulation of hot surfaces.
- No glass equipment used for flammable or hazardous materials, unless no suitable alternative is available.

The principles and general approach used in the Dow method of hazard evaluation have been further developed by ICI Mond Division. The main developments made to the Dow index in the Mond index are:

- It covers a wider range of process and storage installations.
- It covers the processing of chemicals with explosive properties.
- A calculation procedure is included for the evaluation of a toxicity hazards index.
- A procedure is included to allow for the off-setting effects of good design, and control and safety instrumentation.
- The procedure has been extended to cover plant layout.

- Separate indices are calculated to assess the hazards of fire, internal explosion and aerial explosion.

A Hazard and Operability Study (HAZOP) is a procedure for the systematic, critical, examination of a chemical engineering process. When applied to a process design or an operating plant, it indicates potential hazards that may arise from deviations from the intended design conditions. The technique was developed by the Petrochemicals Division of ICI, and is now in general use in the chemical and process industries. A HAZOP study enables a systematic study of the design to be carried out, vessel by vessel, and line by line. Generally, 'Guide Words' are used to help generate thought about the way deviations from the intended operating conditions can cause hazardous situations. These guide words consider, for example, 'too much' as in pressure, temperature or flow rate. A HAZOP study would normally be carried out by a team of experienced people, who have complementary skills and knowledge, led by a team leader who is experienced in the technique. A preliminary study can be made from a description of the process and the process flow-sheets. For a detailed, final study of the design, the flow-sheets, piping and instrument diagrams, equipment specifications and layout drawings would be needed. For batch processes, information on the sequence of operation will also be required, such as that given in operating instructions, logic diagrams and flow charts.

It is clear from this very brief description that the consideration of health, safety and loss prevention has been extensively developed and applied within the chemical engineering discipline. The case studies in Chapter 1 (case studies 1–7) which highlight technical, hardware issues, are attempting to understand incidents against this background of safety engineering.

2.1.2 Human Error

People make mistakes. This is normal and natural and part of the learning process. However, in hazardous situations human failures can give rise to unacceptable consequences such as accidents in the work place. Because of this there is an extensive literature on the subject area of human error.[2-7]

The starting point in attempting to give a brief overview of human performance and human error is to consider human activity in the context of the performance of a task. The concept of an error in performing the task cannot be separated from the idea of a goal. Much human activity is goal orientated and an error can be said to occur when this goal is not achieved.

Types of human performance are often classified as skill-based, (simple almost automatic behaviour patterns) rule-based (diagnosis or actions based on rules: if, then performance) and knowledge-based (performance worked

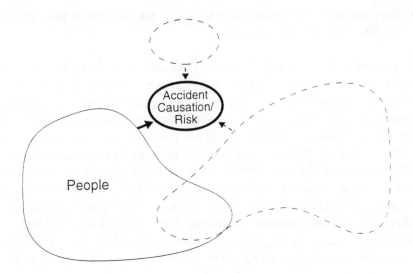

Figure 2.2 *Some models of accident causation emphasise the failures of people and human error as direct causes of accidents, while the failures of hardware or systems and cultures are not emphasised*

out from first principles). These three levels of performance are therefore simply referred to as skill-based, rule-based and knowledge-based.

There are two distinct types of thinking: the conscious or 'attentional mode', which is characteristic of learning and unfamiliar situations, and the automatic or 'schematic mode', characterised by heuristics (rules of thumb) and schemata, i.e. pre-programmed sequences of behaviours, or thought processes, which are invoked by specific conditions or situations.

The three levels of performance, i.e. skill-based, rule-based and knowledge-based, correspond closely to the two distinctive operations of the attentional and schematic modes. The skill-based and knowledge-based levels are readily identifiable with the operations of the schematic and attentional modes, respectively, while the rule-based level could implicate either, i.e. rules may involve conscious knowledge, as in the case of remembered instructions for dealing with an emergency, or they could involve the unconscious application of schematized rules of thumb. When performing a task people will employ the schematic mode in preference to the laborious and resource-limited attentional mode.

The types of error which can occur are related to the types of human performance and types of thinking discussed above. The preference for the schematic mode over the attentional mode leads to skill-based slips or lapses, which occur during the execution of relatively habitual and well intentioned sequences of operations, and to rule-based mistakes which may relate to errors of diagnosis or errors of action. Mistakes at the knowledge-based level also occur. Once a repertoire of schemata and rules is exhausted by the demands of a novel situation, people are thrown back onto their knowledge of the general engineering principles governing a situation. This knowledge is rarely adequate in the case of a complex system.

In addition, the performance of a task is influenced by a wide range of factors which are referred to as performance-shaping (or influencing) factors. The person who performs the task can be considered as an individual, as a member of a team and as part of an organisation. The task is not performed in isolation, but within a working environment. Thus performance-shaping factors can include a wide range of influences from the personal characteristics of the individual, through the levels of stress and comfort under which the task is performed to complex issues such as conflicts of interest, which might arise when safety and output objectives do not correspond. Such conflicts suggest a poor safety culture and may lead to actions that apparently wilfully neglect safety requirements, which are referred to as violations. These comprise a further category of error.[5]

Table 2 attempts to give a summary of the main points, while Figure 2.3 provides another summary of this discussion.

Table 2 *An overview of human error types and their relationship to human performance and thinking*

The way people perform	The way people think	The types of errors people make
Skill-based	Automatic	Slips or lapses
Rule-based		Mistakes of action or diagnosis
Knowledge-based	Conscious	Knowledge-based mistakes

There has clearly been extensive work done to understand human error in a wide range of situations which include hazardous technologies. The case studies in Chapter 1 (case studies 8 and 9) which consider human error as a direct cause of incidents do so against this background. Human error as a direct cause of an incident may be implicated with apportioning blame when an operator makes a mistake, but case studies 8 and 9 also consider the underlying causes of the incidents. This is important because, unfortunately, human error not only has implications of blame and culpability but may also be relevant to questions of negligence and litigation.

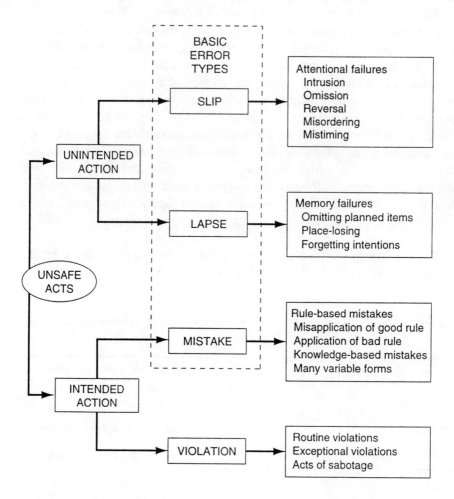

Figure 2.3 *Human error types*
(After Reason, 1990; modified from Ref. 6)

2.1.3 Safety Culture and Attitudes to Safety

The term safety culture was probably first used by The International Atomic Energy Agency (IAEA)[8]. They give the following definition:

Safety culture is that assembly of characteristics and attitudes in organizations and individuals which establishes that, as an overriding priority, nuclear plant safety issues receive the attention warranted by their significance. This statement was carefully composed to emphasize that safety culture is attitudinal as well as structural and relates both to organizations and individuals.

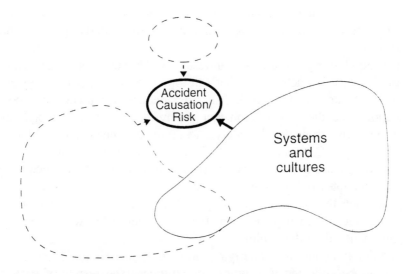

Figure 2.4 *Some models of accident causation emphasise the failures of systems and cultures as the underlying or root causes of accidents.*

The definition relates safety culture to personal attitudes and habits of thought and to the style of organizations. Such matters are generally intangible, but nevertheless lead to tangible manifestations.

The CBI[9] describes the culture of an organisation as the mix of shared values, attitudes and patterns of behaviour that give the organisation its particular character. Put simply it is 'the way we do things round here'. It suggests that the safety culture of an organisation could be described as the ideas and beliefs that all members of the organisation share about risk, accidents and ill health.

A possible shortcoming of the IAEA definition commented on by The Advisory Committee on the Safety of Nuclear Installation Study Group (ACSNI)[4] is that they use the term to describe only an ideal safety culture. ACSNI therefore suggests the following as a working definition:

> The safety culture of an organisation is the product of individual and group values, attitudes, perceptions, competencies, and patterns of behaviour that determine the commitment to, and the style and proficiency of, an organisation's health and safety management.

This definition of safety culture would cover both 'good' and 'poor' safety cultures. Based on the above definitions it seems reasonable to take the view that two important parts of safety culture are the underlying beliefs and attitudes to safety, and their tangible manifestations in the form of safety management systems, levels of competency and so forth (see the following section on safety management systems). There are clearly strong interactions between the tangible and underlying parts of the culture and the term safety culture increasingly emphasises the underlying rather than the tangible.

Thus, good safety culture is manifested by having safety management systems in place in such a way that safe operation of the plant is ensured and accident rates are low. Such issues as adequate resources, good communications and the co-operation that ensures the balancing of pressures which may arise from health, safety and production needs are highlighted and reconciled. Conversely, the lack of good safety culture will be manifested through a weak safety management system. The implication is that accident rates will be high on a site when the safety culture is poor, and low when the safety culture is good. Or, alternatively, that incidents which are allowed to happen when the safety culture is weak would be prevented when the safety culture is strong.

Information about the underlying attitudes and beliefs of the safety culture can be collected via surveys of attitudes. The results of such an attitude survey can be used to construct robust scales of attitudes which correlate with accident performance data for the companies concerned.[10–12]

The ACSNI report[4] is a very comprehensive and authoritative document and develops these themes based on a wide variety of evidence. The Study Group concluded that:

- different organisations doing similar work are known to have different safety records, and certain specific factors in the organisation are related to safety.

For example, in the US nuclear industry, research sponsored by the United States Nuclear Regulatory Commission (USNRC) suggests that good safety performance is predicted by:

- a high level of communication between and within levels of the organisation;
- a mechanism by which the organisation can learn and improve its own methods;
- a strong focus on safety by the organisation and its members;
- some external factors, including the financial health of the organisation and the regulatory climate.

Other evidence is summarised:
Accident rates tend to be lower when:

- resources of time, money and other limited assets are devoted to safety. That is, when there is evidence of strong commitment that is not merely verbal.

- participative relations exist between staff at different levels. That is, all members of staff identify hazards and suggest remedies, provide feedback on the results of action and feel that they 'own' the procedures adopted to pursue safety. There are comprehensive formal and informal communications.
- visibility of senior management on the shop floor is high.
- need for production is properly balanced against safety so that the latter is not compromised.

Quality of training is also associated with a good accident record. The first requirement is training of management, which should include ways of ensuring safety as well as economic efficiency.

Case study 12 from Chapter 1 can be viewed against this background of what constitutes a good safety culture. In case study 12, safety culture issues were considered important because there seemed to be a possibility that safety issues were negated in preference to production pressures and contractual issues.

2.1.4 Safety Management Systems

Safety management is management applied to achieving safety, where safety is taken to be freedom from unacceptable risks that are harmful to people either local to the hazard or elsewhere.

In the HSE publication *Successful Health and Safety Management*[13] (often referred to as HS(G)65), management is described in terms of setting policy; organising and planning; measuring and reviewing performance; and auditing. This is illustrated in Figure 2.5.

HS(G)65 describes the key elements for effective management based upon an analysis of the companies which succeed in achieving good standards. The elements are a mixture of overall management arrangements, individual systems which determine how risks are controlled and practical workplace precautions.

- Workplace precautions provide protection at the point of risk, e.g. safety hardware such as machine guards, protective goggles and pressure relief valves as well as working practices – the way things are done – and control documents such as permit-to-work forms.
- Risk control systems set out the way workplace precautions are implemented and maintained. These may not necessarily be documented systems.
- Management arrangements cover all elements, e.g. policy/purpose, organising, planning and implementing, measuring performance and reviewing and auditing. They are the managerial methods by which an organisation sets out to determine and provide adequate control systems.

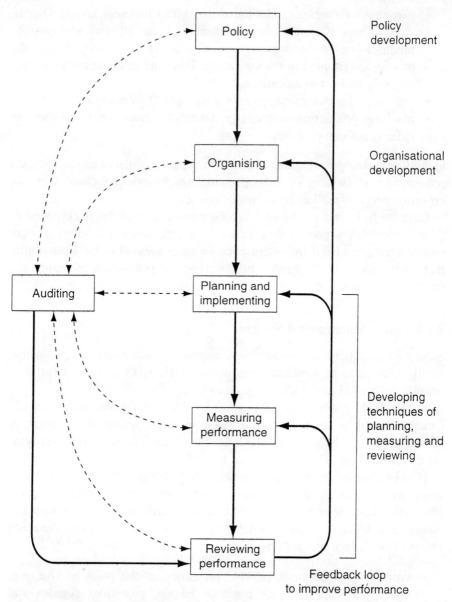

Figure 2.5 *Safety management and the management control loop as shown in Successful Health and Safety Management (HS(G)65)*
(Modified from Ref. 13)

In the best companies management arrangements, risk control systems and workplace precautions support each other. They form a logical structure resulting in effective control of risk. There should be risk control systems for each significant hazard to which either employees or members of the public are exposed. They will ensure that the correct workplace precautions are in place and maintained. To be effective, each risk control system should consist of the management activities outlined in HS(G)65, and hence will consist of complete systems in their own right.

Clearly the total safety management system on a site is a complex set of interacting documents, procedures and concepts, which control all the risks/hazards on the site. Case studies 10 and 11, listed in Table 1 as 'failures of management', can be viewed against this description of successful safety management. The case studies show that the management systems were clearly deficient or defective in certain ways, e.g. systems did not exist or were very weak. As a result of the failures of management there was a lack of control of hazards and risks at the 'sharp end' of the business concerned.

2.1.5 Sociotechnical Systems

Sociotechnical systems are systems which have both human and technical components. Clearly the theory of such systems is highly relevant to understanding accident causation in a variety of situations including industrial plants and transport systems. An article by Bill Richardson[14] has examined the profile and prevalence of sociotechnical disasters. Such disasters occur when human, organisational and technical systems, in combination, break down. They usually incur massive economic and social costs and create large-scale damage. The approach emphasises the multiple causality of accidents and the concept of immediate (or direct) causes and underlying (or root) causes of accidents.

Another approach is the systems failure method[15,16] which analyses failures of sociotechnical systems. The method is very flexible and can examine all types of systems failure at a variety of levels of detail. It incorporates the concept of 'mismatch' between the ideal system (called the 'system paradigm') and the actual system observed in practice. When this comparison is made, the causes of failure in the system may be attributed to such issues as:

- deficiencies in the apparent organisational structure such as lack of a performance-measuring subsystem.
- no clear statements of purpose supplied in a comprehensible form from the wider system.
- the various subsystems not having an effective means of communication.
- a subsystem being inadequately designed.
- insufficient consideration having been given to the influence of the external environment so that sufficient resources to cope with foreseeable environmental disturbances had not been built into the organisation.

The system 'paradigm' is part of a formal method (the failures method) to analyse the failure of sociotechnical systems. Other aspects of the method consider failure in other important areas such as:

- Control
- Communication
- Engineering reliability
- Human factors

The method of comparison between the actual system and a 'paradigm' for that area of investigation (e.g. human factors) is used throughout. In this manner a broad based and systematic method is produced which can identify many different direct and underlying causes for accidents at many different levels within an organisation or beyond it, i.e. in the environment of the system or relating to the context in which the organisation operates.

2.1.6 Latent Failures and Active Failures

The concept described above of multiple causality of accidents and incidents has been extensively developed by Reason,[6,17] who draws a distinction between active and latent failures (see also Section 2.1.2 on human error). Active failures are generally made by front-line operators, e.g. pilots, drivers, control room staff and machine operators. Latent failures are made by those whose activities are removed in time and space from operational activity – designers, decision-makers and managers. Latent failures would appear to be equivalent to the underlying causes of failures, while direct causes are equivalent to active failures.

Reason also proposes a resident pathogen metaphor: latent failures are analogous to resident pathogens in the body which combine with external factors to bring about disease. For example, in giving evidence to the House of Lords[18] he says:

> Rather than being the main instigators of an accident, operators tend to be the inheritors of 'pathogens' created by poor design, incorrect installation, faulty maintenance, inadequate procedures and management decisions, and the like. The operators' part is usually that of adding the final garnish to a lethal brew that has been long in the cooking. In short: unsafe acts in the 'front-line' stem in large measure from bad decisions made by the rear echelons.

There are clear parallels with the sociotechnical failures method which also considers a broad range of direct and underlying accident causes.

Many other comprehensive accident causation models exist. Mention should be made of the MORT approach;[19] the ILCI loss causation model[20] the Hale and Glendon causation model[21] and the STEP procedure.[22] Ferry[23] provides a review and the CCPS[7] gives a good summary.

This is not a comprehensive review of the failures method or of accident causation models. However, it is clear that consideration of accident causation in a broad based, systemic or holistic manner shows that most accidents are characterised by:

- Multiple causes. (Some are direct causes, others are underlying or root causes.)
- Human errors and failures of hardware that are often the last 'symptom' of poor design, poor procedures or poor training, for example. These in turn result from poor definition of tasks, conflicting requirements and inadequate resources.

2.1.7 High Reliability Theory and Normal Accidents

In his book *The Limits of Safety* Sagan[24] discusses two approaches (schools of thought) to considering the origin of accidents. These are based in two different approaches to the theory of organisations.

- High reliability organisation theory.
- Normal accident theory.

To quote Sagan:

> High reliability theorists believe that hazardous technologies can be safely controlled by complex organizations if wise design and management techniques are followed. This optimistic conclusion is based on the argument that effective organizations can meet the following four specific conditions, which are necessary to create and maintain adequate safety:
>
> - political elites and organization leaders place a high priority on safety and reliability;
> - significant levels of redundancy exist, permitting backup or overlapping units to compensate for failures;
> - error rates are reduced through decentralization of authority, strong organizational culture, and continuous operations and training; and
> - organizational learning takes place through a trial-and-error process, supplemented by anticipation and simulation.

These conditions have been witnessed in a number of high reliability organisations, and if these conditions exist in other organisations, then the theory would predict that serious accidents and catastrophes can be prevented. As Macrone and Woodhouse [Ref. 25] put it, "While the exact mix of strategies appropriate in a given case obviously depends on the nature of the particular problem, the catastrophe-aversion strategy outlined above should be applicable to virtually any risky technology." Others [Ref. 26] hold similar views: ' ... most of the characteristics identified here should operate in most organizations that require advanced technologies and in which the cost of error is so great that it needs to be avoided altogether'. Thus, while the high reliability theorists do not state what precise amounts and mixtures of these factors are necessary for operational success with hazardous technologies, their overall optimism is clear. Properly designed and well-managed organizations can safely operate even the most hazardous technologies.

However, this has to be contrasted with Normal accident theory.[27]
From Sagan again:

> Normal accidents theorists take a view in which organizations and members of organizations are self-interested actors with potentially conflicting interests, and in which organizations are strongly influenced by broader political and social forces in the environment. These theorists view organizations as often having inconsistent preferences, unclear technologies, and fluid participation. The theory predicts that serious accidents are inevitable if the organizations that control hazardous technologies display both high interactive complexity and tight coupling.

Sagan explains that interactive complexity is a measure, not of a system's overall size or the number of subunits that exist in it, but rather of the way in which parts are connected and interact. According to Perrow[27] complex interactions are those of unfamiliar sequences, unplanned and unexpected sequences that are either not visible or not immediately comprehensible. Whether a system is tightly coupled or loosely coupled affects its ability to recover from small-scale failures before these cascade into larger problems. Tightly coupled systems have more time-dependent processes; the sequences and co-ordinated activities needed to produce the product are invariant and have little slack, i.e. quantities used in production must be precise and the process must be done right the first time or not at all.

Sagan continues:

> Each of the four factors previously identified as contributing to high reliability are seen from the normal accidents' perspective as being ineffective, unlikely to be implemented, or even counter-productive. Even if leaders place a very high priority on safety and reliability, which is by no means a given, competing organizational and individual objectives will remain: the continuing desires to maximise production, maintain autonomy, and protect personal reputations, however, can severely impair efforts to improve safety. Adding redundancy does not necessarily enhance reliability because it also increases interactive complexity, encourages operators to run more risks, and makes the overall system more opaque. Decentralized

decision-making does not necessarily enhance safety because tightly coupled systems require rapid responses and strict adherence to standard operating procedures. Intense socialization and a strong organizational culture are unlikely to be very productive in hazardous organizations both because their leaders cannot know how operators should respond in all contingencies and because democratic societies are unwilling to isolate and control all aspects of the lives of such organizations' members. Constant training and practice will not address accident scenarios that are unanticipated, excessively dangerous, or politically unpalatable.

Finally, a number of factors will severely constrain the process of trial-and-error learning: uncertainty about the causes of accidents, the political interests and biases of organizational leaders and low-level operators, and compartmentalization within the organization and secrecy between organizations.

These two views are further discussed in Section 2.2 below.

2.1.8 The Subjective/Objective Debate

There is a long-standing debate in the social sciences as to the most appropriate philosophical position, or tradition, from which to investigate the social world[28] (which could include a major hazard chemical plant). The poles of this debate are defined by the traditions of positivism on the one hand and phenomenology on the other.

- The key idea of positivism is that the social world exists externally and is objective and 'real'.
- The key idea of phenomenology is that the world, or 'reality', is not objective and exterior but are socially constructed and given meaning by people. The world can be known only subjectively. This view is also referred to as relativism, and is currently labelled social, cultural or 'postmodernist' relativism.

There are of course other traditions which can be placed on a continuum between extreme subjectivist and extreme objective approaches.[28]

It is important to be aware of these traditions when considering risk because there are parallel positions to this in the risk debate, represented by extremes which describe risk as an objective and measurable quantity or as an entirely subjective and socially constructed concept. As Clive Smallman[29] puts it:

> Perhaps chief amongst the problems facing researchers in risk (and those who must manage it) is that risk itself is an abstract phenomenon: 'the probability that a particular adverse event occurs during a stated period of time, or results from a particular challenge' [Ref. 30].

Hazard can be objectively identified as can its aftermath, but risk depends on a complex interplay of social variables, that are ultimately combined by human judgement (this is the subjective or socially constructed view). Yet, many insurers, risk assessors, engineers and other professionals (whose very livelihood depends on risk) insist that this abstract socially-formed and singularly subjective phenomenon may still be objectively measured (this is the objective or positivist position). In this case public attitudes and beliefs (perceptions) are devalued in favour of the 'expert' view (supposed 'reality'). Paradoxically, writers such as Johnson and Covello [Ref. 31] propose that it is a rich socially constructed reality that is substituted by abstract rationality – the 'expert's reality'. There are clear parallels with the debate that exists between the phenomenologist and positivist schools in social research.

It is interesting to note that the objective/subjective (or positivist/social construction) debate seems to centre around the idea of 'risk' not the idea of 'hazard', i.e. it is commonly accepted that hazards, and the realisation of hazards, are 'objective', but that risk is a much more subjective and constructed idea involving much human judgement. It may be useful to add some further observations at this point.

- Risk is not a uniform idea. The risk of 'say' dying in a road accident is 'more objective' than the risk of a major chemical accident occurring.
- Hazards are objective reality. Real people get hurt, or are killed, by real hazards at work every day.
- Activities using similar technology nevertheless pose different risks to the people involved with them.[10, 4] Thus despite the uncertainty about estimating risk it is true that two individuals doing similar jobs in different places will run different risks. Because of this the management of risk is an important issue.[10]

This idea is powerfully developed by Marcelo Firpo de Souza Porto and Carlos Machado de Freitas in their work on the socio-political amplification of risk[32] (NOT the social amplification of the perception of risk; see Glossary). They say:

Our main objective is to examine the increasing risk of industrializing countries. It is possible that, in attempting to detect the causes for major chemical accidents, the overriding factor to be considered should be the characteristics of the context (or environment) where such accidents take place . . .

The point they are making is that people working in industrialising countries have a greater risk, for a given technology, than people in first world countries.

Similarly, Pier-Alberto Bertazzi[33] says:

> The characteristics of the release context (of a toxic substance) might be as important as the nature of the released materials in determining exposure, and they might well override the absolute magnitude of the release in relevance. Bhopal investigations, for example, shed light on several off-plant determinants of the disaster. Some of them were obvious, such as the siting of an urban aggregate of several hundred thousand people in the immediate surroundings of the plant. Other determinants were housing structure, height from the ground, the narrowness and poor illumination of the streets, the lack of communication and transportation, the lack of knowledge and warning of the incumbent hazard, etc.

A similar general point is made by Shivastava[34] considering the nature of industrial crises. He says:

> The role of the context or environment is de-emphasized in most explanations of technological accidents. This analysis shows that contextual or environmental variables set up the preconditions for accidents, cause triggering events, and also exacerbate the effects of accidents after they occur. A common form of interaction in which failures occur is communication between organizational decision makers and outsiders.

The work of ACSNI[4] also points to the important conclusion that different organisations doing similar work are known to have different safety records. It is therefore clear that both organisational factors and the nature of the context and the environment in which organisations exist and operate are important factors in determining the accident rates and levels of safety at different industrial plants. These are clearly objective issues. The subjective/objective debate is further discussed throughout this book.

2.1.9 Risk Perception and Acceptable Risk

The objective/subjective debate discussed above provides an important background to the concepts of risk perception and acceptable risk.

A Report from a Royal Society Study Group[30] includes a chapter on risk perception. Four major themes from the recent literature are identified.

- The view that a separation can be maintained between 'objective' or 'actual' risk and 'subjective' or 'perceived' risk has been largely abandoned. Most people would agree that the physical consequences of hazards, such as deaths, injuries and environmental harm, are objective facts. However, assessments of risk, whether they are based upon individual attitudes, the wider beliefs within a culture or the

models of mathematical risk assessment, necessarily depend upon human judgement. In this respect it can be argued that assessments of risk involve a degree of subjectivity, to a greater or lesser extent.
- The important work by Fischhoff et al.[35] on acceptable risk has been broadly accepted. In this view acceptable risk problems are decision problems, i.e. the best way of viewing acceptable risk is as a decision problem. It is not a particular risk which is acceptable, or not. It is a particular decision which is acceptable, or not, to certain people involved. Therefore, there can be no single, all-purpose number that expresses the acceptable risk for a society. It is necessary to consider risk to whom, when, and in what circumstances. It is important to consider the choices people have in making decisions about risk.
- Many researchers have sought to look beyond purely individual, psychological explanations of human responses to hazards. Social, cultural and political processes are now acknowledged as all being involved in the formation of individual attitudes towards risks and their acceptance. This broadening of horizons has involved a higher profile for sociologists and anthropologists within the field of risk studies, together with attempts to adopt genuinely interdisciplinary approaches to this issue.[35, 36]
- The emergence of risk communication as a topic of concern. This involves the key question of public involvement in decision making about hazards and risks.

2.2 AN ANALYSIS OF DIFFERENT APPROACHES

2.2.1 Accident Causation

Some synthesis can be achieved between some of the strongest ideas which have been discussed above especially in Sections 2.1.3–2.1.7 which relate to issues of systems and culture as underlying causes of accidents and incidents. Let us consider high reliability organisations.[24]

The main characteristics for high reliability organisations are:

- commitment to safety;
- redundancy in both engineering systems and human systems;
- decentralisation, strong cultural commitment to safety and continuity of the work force;
- oganisational learning.

These characteristics would seem to be similar to those which are described for organisations which have a strong safety culture. For example the work of the ACSNI committee on human factors[4] summarises characteristics for strong safety cultures. These characteristics are:

- senior management are strongly committed to safety and have a high level of visibility on safety matters;
- strong communications exist between senior management and the workforce and within the workforce;
- the idea of safety and production trade-offs is accepted, as is the need to show a high level of commitment to safety as opposed to production;
- the standard of competence and of training of the workforce and management is high;
- the workforce is older and more experienced.

The ACSNI work also lists characteristics which predict good safety performance (see Section 2.1.3). These characteristics include a mechanism by which the organisation can learn.

Thus, lists of characteristics for high reliability organisations and organisations with a strong safety culture would seem to overlap strongly. They are not identical descriptions but the view can be taken that high reliability theory and strong safety culture theory are drawing from similar sources.

Similarly, organisations which achieve high standards of safety management[13] are described in terms of commitment, good communications, clear policy objectives and organisational learning. These characteristics seem to be strongly related to the ideas of strong safety culture and high reliability.

When failure is analysed using the systems failure method[15, 16] a system 'paradigm' is used (Section 2.1.5). The description of a system 'paradigm' would include such 'ideal' features as:

- a rational system with appropriate subsystems for decision making, performance monitoring and rational decision making.
- the existence of a wider system within which the system operates. This wider system or context formulates the initial design of the system, provides resources for it and formulates the exceptions to the systems under consideration.
- effective communications between the subsystems, and between the system and its wider environment.

This description of a system paradigm is again similar to the description of a high reliability organisation, implying highly rational and well designed organisations. In other words we can say that the system paradigm of an organisation, the high reliability organisation description, the successful safety management description and the description of an organisation with a strong safety culture have strong links and similarities with each other and essentially represent some sort of 'ideal' situation in which the organisation can run with high reliability despite having to deal with highly hazardous technologies.

In the systems failure method[15,16] the approach which is used when considering any system failure is to construct the 'real' system which has been observed to fail and to compare it with the system paradigm as outlined above. Mismatches or differences between the real system and the system paradigm are used as ways of indicating weaknesses in the real system. These weaknesses in the real system are those which have the potential to give rise to failures within the system, i.e. erode the reliability of the system.

An alternative theory of organisational functioning which is also less optimistic about reliability is provided by the model of the normal accident theorists.[27] This approach describes what many would recognise from their own experience as a 'real' organisation. The type of organisation described by normal accident theory shows:

- excessive effects due to the influence of the context or environment;
- interactive complexity within the organisation which makes it hard to understand and so very prone to what reliability engineers call common mode failure;
- tight coupling creating pressure within the production processes and
- conflicting interests between key members of the organisation or between the organisation and its environment.

The way the systems failure method seeks to model real systems is similar to the normal accident framework, so that we can make a synthesis between the system description of a real organisation in the failure method and the descriptions of organisations which give rise to normal accidents.

What this means is that there are four approaches to describing 'ideal' organisational structures for achieving higher reliability.

- High reliability organisation theory.
- Organisation theory which points to strong safety culture.
- The system failure approach which describes a system paradigm.
- Descriptions of successful health and safety management.

We might take these four areas of description to be essentially equivalent to each other or at least similar. On the other hand there are descriptions of 'real' organisations where this kind of 'idealness' or paradigm is not in practice achieved. This is what is talked about in normal accident theory and this is what is described and constructed in the systems failure method which seeks differences between the paradigm and the real situation.

The approaches of the normal accident theorists and the high reliability theorists are sometimes set against each other in the sense that these schools of thought are supposed to be alternative descriptions of organisations with either one being true or the other being true. This does not seem to be a helpful way of interpreting the work which is done by the respective groups. I would take the view that there are 'ideals' or 'paradigms' of high reliability organisations, or organisations with strong safety cultures which can lead to more reliable operation. However, in most organisations things are not ideal. In fact often real organisations do have conflicting interests, real organisations hide information, real organisations have problems of how they appear to the external world and how they can manage that appearance. Therefore, what exists in practice is a set of organisational characteristics which can be set against the 'paradigms' of 'idealness'. The deviations from these ideals are what drives the faults and problems within real organisations.

These faults and problems are also referred to by such workers as Reason[6] and called latent failures. Quite clearly latent failures (which relate to problems such as poor definition of organisational goals, poor organisational design, unclearness of safety objectives and so forth) undermine the reliability of organisations and can be described in terms of high reliability theories not working properly, strong safety culture not working properly, system paradigms breaking down, or safety management systems being inadequate. This discussion is illustrated in Fig 2.6.

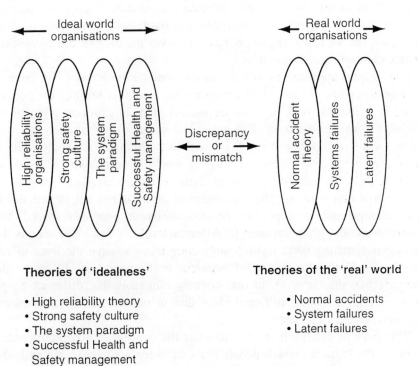

Figure 2.6 *'Ideal' and 'real world' descriptions of organisations. The figure shows that four descriptions of 'ideal' organisations are broadly similar, while three descriptions of 'real world' organisations also show commonalities. The discrepancy between the 'ideal' world organisations and the 'real' world organisations are what drives the faults and problems within real organisations*

Note also that there is an implication that 'ideal' organisations have good accident performance while 'real' organisations may have poor accident performance (high accident rates). The greater the discrepancy between the real world organisation and the ideal the greater the accident rate difference is likely to be.

2.2.2 Subjective/Objective

This section attempts to develop further the concepts introduced in Sections 2.1.8 and 2.1.9, which described the subjective/objective debate and issues of risk perception and acceptable risk. Otway and Thomas[37] have described this debate in risk assessment circles as being essentially reducible to whether or not one has an ideological commitment either to an objective and ultimately assessable science of facts (the positivist view) or to an essentially subjective and ultimately socially constructed view of the world. The discussion of the subjective and ultimately constructed view of the world is described in the cultural theory of risk,[36] whereas the descriptions of an objective and ultimately assessable science of facts are more to do with engineering reliability and hardware assessments of risk.[1] However, the idea that engineers have no understanding of the complexity of risk concepts seems doubtful.

One helpful approach to this dichotomy may be to consider risk from an experiential point of view.[32–34] In other words, what is life like for people? What is it like to work on a major hazard plant in different parts of the world? Or what is it like to work in a plant where safety is taken seriously as opposed to one where it isn't? Is there such a thing as the socio-political enhancement of risk (Section 2.1.8)? It would appear from the evidence from major hazard disasters such as those which occurred at Mexico City and Bhopal that whatever the difficulties of defining risk might actually be risk management is important because different people involved with similar technology are exposed to different levels of risk. This means that getting risk management right is something that can save the lives of real people. Similarly, the range of accident performance of different sites carrying out similar work in one country illustrates that different people doing similar jobs run different risks due to contextural differences and organisational differences.

The work of Fischhoff et al.[35] looks at the concept of acceptable risk in terms of the balances which people make between what they gain and what they lose in taking particular risks. This is why under different circumstances people make different decisions or choices about what is an acceptable risk to them. This seems to be a perfectly legitimate and sensible, human-based, approach to the kind of decisions that people make about risk. One of the great problems of making risk-based decisions is that the people who often make the decision, and the people who actually benefit from the decision are not the people who have to bear the consequences of those risks. Under these circumstances there exists a dichotomy between the two sides of the equation, i.e. who gains and who loses, and this is one reason why the concept of acceptable risk may be difficult to define in numerical terms. Thus Fischhoff explains that under any set of circumstances people will balance the risk they take against the benefit they draw. Let us think about a few examples: if people benefit from the thrills of motor cycle racing, hang-gliding or rock climbing, if those activities are extremely exciting and extremely stimulating to the individuals concerned, then they will quite clearly choose to bear the risk in order to take part in those activities. However, if people are told that a land-fill site is going to be built near their houses for disposing of waste there will be huge local opposition to that, because the local people are on the losing end of a community decision which may on balance seem right to the people taking the decision but which the local people (on the receiving end of the decision) do not see as right at all. Comparing the risks of land-fill sites with those of hang-gliding is not helpful in this situation. There is no universal risk level which is 'acceptable' in all circumstances.

There is a large literature on risk perception. This can be used to support a view that there are right and wrong values for risk and right and wrong decisions in risk-based decision making. The work on risk perception shows how different factors enter into the decision-making arena for individuals and is very useful, but it will never get around the fact that different people are going to have legitimate and different perspectives in any risk-based decision situation. The only way that these different perspectives can be reconciled is by some process of communication. Thus although the risk perception work is valuable, it clearly sets out a whole series of factors which affects people's perceptions of risk, and clearly people's ability (and this includes experts) to judge risk and to rank risks sensibly in quantitative order is limited. However, that in no sense establishes that people's ability to make rational decisions about risk is impaired. Thus risk communication is a highly complex issue in its own right and can only be useful if it is based on trust and mutual respect between the various people involved in a risk-based decision. Attempts to use arguments based on acceptable risk levels will not succeed in situations in which some people are on the 'losing' end of the decision making process.

2.3 SUMMARY AND DISCUSSION

This chapter has very briefly reviewed the different theoretical approaches which have developed to provide a better understanding of industrial accidents and incidents from the perspectives of safety engineering, human error and systems and cultures. These perspectives were introduced in Chapter 1 by the use of simple case studies of accident causation and investigation. In addition the complexity of risk as a concept has been introduced under the broad heading of a subjective/objective debate and the discussion has included important ideas related to risk perception, risk communication and acceptable risk.

It is clear that the consideration of health, safety and loss prevention has been extensively developed and applied within the chemical engineering discipline. The case studies in Chapter 1 (case studies 1–7) which highlight technical, hardware issues are, therefore, attempting to understand incidents against a background of safety engineering.

In addition there has been extensive work done to understand human error in a wide range of situations which include hazardous technologies. The case studies in Chapter 1 (case studies 8 and 9) which consider human error as a direct cause of incidents do so against this background.

The approaches described under the heading of systems and cultures and which are illustrated in Chapter 1 (case studies 10–12) show that consideration of accident causation in a broad-based systemic or holistic manner indicates that most accidents are characterised by:

- Multiple causes. (Some are direct causes; others are underlying or root causes.)
- Human errors and failures of hardware that are often the last 'symptom' of poor design, poor procedures, or poor training, for example. These in turn result from poor definition of tasks, conflicting requirements and inadequate resources.

In considering organisational factors (issues of systems and cultures; Sections 2.1.3–2.1.9) some simple synthesis is suggested in the chapter between some of the approaches described, for example:

- high reliability organisation theory
- strong safety culture descriptions
- a system paradigm description
- descriptions of successful health and safety management.

These show some commonalities and overlap and may represent 'ideal' organisations (Figure 2.6).

On the other hand 'real world' descriptions are included in the ideas of latent failures, the systems failure method and normal accident theory. The implication is that accident rates will be higher in organisations which are significantly different from the ideal. It is clear that both organisational factors and the nature of the context and the environment in which organisations exist and operate are important factors in determining the accident rates and levels of safety of different industrial plants. It is suggested that the differences or discrepancies between the 'ideal' world organisations and the 'real' world organisations are what drives the faults and problems within real organisations.

The subjective/objective debate shows the complexity of risk as a concept and shows that some people consider risk to be objective and real while others consider risk as a subjective concept. Nevertheless some important observations can be made.

- Risk is not a uniform idea. The risk of dying in a road accident is 'more objective' than the risk of a major chemical incident occurring.
- Hazards are objective reality. Real people get hurt, or are killed, by real hazards at work every day.
- Activities using similar technology nevertheless pose different risks to the people involved with them.[10, 4] Thus despite the uncertainty about estimating risk it is true that two individuals doing similar jobs in different places will run different risks. Because of this the assessment and management of risk is an important issue.[10]

I express caution about the use of the literature on risk perception when it is used in an attempt to establish that there are right and wrong decisions to be made about acceptable risk, especially when this is based on the notion of an objective, or real, risk as opposed to a subjective, or perceived, risk. This can in turn lead to the notion of risk communication as a way of 'explaining' that certain risks are 'acceptable'. This approach is at variance with an extensive and well considered literature on acceptable risk as a decision problem. Acceptable risk depends on risk to whom and in what circumstances. In addition there are real differences in need and outlook of individuals and in some circumstances an acceptable risk may be hard to define at all.

Further Reading

1. *J. Contigencies and Crisis Management*, 1994, **2** (2–4); 1995, **3** (3); 1996, **4** (2).
2. C. Perin, *Industrial & Environmental Crisis Quarterly*, 1995, **9** (2), p. 152.
3. F. C. Lees, Loss Prevention in the Process Industries, Vols 1–3, Butterworth-Heinemann, London, 2nd edn., 1996.
4. O. Renn et al., *J. Social Issues* 1992, **48** (4), 137.
5. G. Burrell and G. Morgan, *Sociological Paradigms and Organisational Analysis*, Gower Publishing, Aldershot, UK, 1985.

References

1. R.K. Sinnott, Coulson and Richardson's Chemical Engineering, Vol 6, Pergamon, Oxford, 1993.
2. I. Glendon and E.F. McKenna, *Human Safety and Risk Management*, Chapman & Hall, London, 1995.
3. Advisory Committee on the Safety of Nuclear Installations, *Second Report: Human Reliability Assessment – A Critical Overview*, HMSO, London, 1991.
4. Advisory Committee on the Safety of Nuclear Installations, *Third Report: Organising for Safety*, HMSO, London, 1993.
5. Health and Safety Executive, *Improving compliance with safety procedures: Reducing industrial violation*, HMSO, London, 1995.
6. J. Reason, *Human Error*, Cambridge University Press, Cambridge, 1990.
7. *Guidelines for Preventing Human Error in Process Safety*, CCPS AIChemE. New York 1994.
8. International Atomic Energy Agency, Safety Culture, Safety Series No.75 – INSAG 4.
9. Confederation of British Industry, *Developing a Safety Culture – Business for Safety*, CBI, London, 1990.
10. N. W. Hurst, J. Loss Prev. Process Ind., 1996, **9**, 161.
11. I. Donald and D. Canter, J. Loss Prev. Process Ind., 1994, **7**, 203.
12. S. Cox and T. Cox, *Work and Stress*, 1991, **5**, 93.
13. Health and Safety Executive, *Successful Health and Safety Management* (HS(G)65), HMSO, London, 1991.
14. B. Richardson, *Disaster Prevention and Management*, 1994, **3**, 61.
15. V. Bignell and J. Fortune, *Understanding Systems Failures*, Manchester University Press, Manchester, 1984.
16. J. Fortune and G. Peters, *Learning from Failure – The Systems Approach*, J. Wiley, London, 1995.
17. J. Reason, Phil. Trans. Roy. Soc., 1990, (**8**) 327, 475.
18. J. Reason, *Human Factors in Nuclear Power Operations*, in House of Lords Select Committee on Science and Technology (Subcommittee II), Research and Development in Nuclear Power, Vol.2 – Evidence, House of Lords paper 14-II, HMSO, London, 1989, p. 238.
19. W.G. Johnson, *MORT Safety Assurance Systems*, National Safety Council of America, Chicago, IL, 1980.
20. F.E. Bird and G.L. Germain, *Practical Loss Control Leadership*, Inst. Publ. Loganville, GA, 1988.
21. A.R. Hale and A.I. Glendon, *Individual Behaviour in the Control of Danger*, Industrial Safety Series – 2, Elsevier, Amsterdam, 1987.
22. K. Hendrick and L. Benner Jr., *Investigating Accidents with STEP*, Marcel Dekker, New York, 1987.
23. T.S. Ferry, *Modern Accident Investigation and Analysis*, J. Wiley, New York, 2nd edn., 1988.
24. S.D. Sagan, *The Limits of Safety. Organisations, Accidents and Nuclear Weapons*, Princeton University Press, Princeton, NJ, 1993.
25. J.G. Macrone and E.J. Woodhouse, Averting Catastrophes P.136, University of California Press, Berkeley, 1986.
26. K.H. Roberts, Some Characteristics of One Type of High Reliability Organization P.173 Organisation Science, 1990, **1**.
27. C. Perrow, *Normal Accidents: Living with High-risk Technologies,* Basic Books, New York, 1994.

28. M. Easterby-Smith, R. Thorpe and A. Lowe, *Management Research – An Introduction*, Sage, London, 1991.
29. C. Smallman, *Disaster Prevention and Management*, 1996, **5**, 12.
30. Royal Society, *Risk: Analysis, Perception and Management*, Report of a Royal Society Study Group, Royal Society, London, 1992.
31. B.B. Johnson and V.T. Covello (eds), *The Social and Cultural Construction of Risk*, Reidel, Dordrecht, 1987.
32. M.F. de Souza Porto and C. M. de Freitas, *Risk Analysis*, 1996, **16**, 19.
33. P.A. Bertazzi, *Scand. J. Work Environ. Health*, 1989, **15**, 85.
34. P. Shivastava, I. Mitroff, D. Miller and A. Miglani, *J. Management Studies*, **25**, 285.
35. B. Fischhoff, S. Lichtenstein, P. Slovic, S.L. Derby and R.L. Keeney, *Acceptable Risk*, Cambridge University Press, Cambridge, 1981.
36. M. Douglas and A. Wildavsky, *Risk and Culture: An Essay on the Selection of Technological and Environmental Dangers*, University of California Press, Berkeley, 1982.
37. H. Otway and K. Thomas, Risk Analysis, 1982, **2**, 69.

CHAPTER 3

The Assessment of Risk – Quantification

Whenever risk is quantified, there is an underlying assumption that the method used to estimate the risk is 'correct' and complete so that it represents all important aspects of the system it purports to model. Thus a quantified risk assessment of a chemical plant, involving hardware failures, modelling of fires and explosions, etc., is taken to be a good model of the system and hence to generate realistic estimated values for the risk. It is also assumed that appropriate data are available for use within the models. The first two chapters of this book have illustrated the different perspectives that may be taken in considering accident causation, and surveyed a range of theoretical approaches which reflect these different perspectives. Furthermore, Chapter 2 has illustrated the complexity of understanding risk (is risk objective or subjective?) and of evaluating such issues as acceptable risk. This complexity is not an issue concerning the inability of some people to understand what is clear to others. The complexity arises from the difficulty of understanding interlinked ideas and thoughts where different views, knowledge and perspectives have legitimate parts to play.

Many workers in the risk assessment field are aware of these different perspectives and complexities. Thus, many accept that a risk estimate which considers only hardware failures and ignores safety management, safety culture and human error cannot be considered complete. Because of this, links between safety cultures, human errors, safety management and risk assessment are important to establish, and this chapter will describe in general terms the efforts which have been made to develop tools which make these connections explicit.

The risk assessment perspectives described in this chapter build on the perspectives of accident causation described in Chapter 1 and the theoretical ideas described in Chapter 2. To illustrate this change of emphasis from an understanding of accident causation to an assessment of risk the figures in Chapter 3 use 'Risk assessment' as their central idea. Hand in hand with this change, there is a tendency for the language of

Chapter 3 to be somewhat more technical than in the previous or subsequent chapters.

3.1 ENGINEERING APPROACHES TO RISK ASSESSMENT

3.1.1 Hazard Identification and Risk Assessment

As explained in the Glossary, the terminology associated with the terms 'hazard' and 'risk' are a particular source of difficulty. To reiterate, a hazard is defined here as an object or a situation with the potential to cause harm. Thus a hazard in the workplace might be a chemical or a machine – anything which can hurt people if certain circumstances prevail, e.g. a chemical is spilt and inhaled, or contact is made between a person and a moving machine, is a hazard. The concept of risk tries to go beyond hazard and to answer the questions (1) how likely is something to go wrong? and (2) what will the effect be? So the definition of risk contains the two elements of frequency (or probability) and consequence.

Risk, therefore, describes both the likelihood of an event and the consequence. But in order to compare risks it is necessary to 'fix' the consequence. For example, if the consequence considered is 'death to an individual' then the risks from a series of activities can be compared by considering the frequency with which the activities cause death. Similarly, if the consequence chosen is an injury resulting in (say) three days lost from work then risks for a series of activities can be compared by considering the frequency with which the activities cause a 'three-days-lost' accident.

However, if one activity is likely to cause death (e.g. by entrapment in a machine) but the likelihood is low, and another activity is likely to cause a less serious injury (e.g. a three-days-lost accident) but with a higher frequency, then no simple and unambiguous method is available to compare the two risks. For chemical process hazards, quantified risk assessments commonly estimate two risk measures to provide an insight into these two dimensions of frequency and severity. Individual risk is the frequency at which an individual may be expected to sustain a given level of harm (e.g. death) from the realisation of specified hazards. Societal risk is the relationship between frequency and the number of people in a given population suffering from a specified level of harm from the realisation of specified hazards.

The Institution of Chemical Engineers publication *Risk Assessment in the Process Industries*[1] is an excellent example of the engineering approach to risk assessment. This work describes the process of risk assessment under the headings:

- Hazard identification procedures
- Consequence analysis

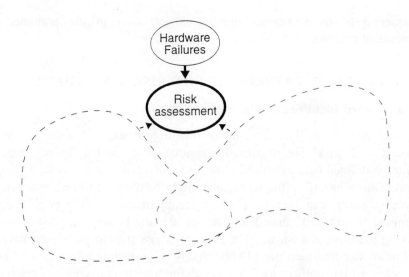

Figure 3.1 *Quantified risk assessment may be based on an analysis of hardware failures and their consequences. The effects of people failures and failures of systems and cultures may not be fully modelled or emphasised*

- Quantification of event probabilities and risk
- The application of risk assessment
- Special topics

The aim of risk assessment is described as providing answers to the questions:

- What can go wrong?
- What are the consequences and effects and are these acceptable?
- Are the safeguards and controls adequate to render the risk acceptable?

Hazard identification procedures include such methods as HAZOP, and the analysis of the steps involved in an operating procedure (referred to as task analysis). These methods often use a team of people with individual skills and experience which are relevant to the type of process and equipment to be reviewed. Although simple in concept, HAZOP has proved very effective over the years. The aim is to identify potential hazards such as fire, explosion, loss of containment and so forth as a result of equipment failure or human error (see Chapter 2).

Consequence analysis is that part of risk assessment which considers the physical effects of hazards and the damage caused by them. For example, a pressurised leak of a flammable substance may give rise to a jet fire, flash fire or vapour cloud explosion. Mathematical models exist to calculate the effects of these events in terms of overpressure and thermal radiation.

Separate mathematical relationships may be used to estimate the effects of this radiation or overpressure on people.

Quantification of event probabilities may involve the use of statistical data on failure of plant items (or whole systems) based on historical experience, or on the estimation of event probabilities using the analytical methods of reliability engineering such as fault tree analysis. In addition, failure rate data derived from both historical and theoretical data, and which may include elements of engineering judgement, may be represented by a single (average) value referred to as generic failure rate data.

Fault tree analysis is the most widely used method for developing failure logic. The process used is to select an undesired 'top event' (e.g. loss of containment of a toxic substance) and trace it back to the possible causes which can be component failures or human errors. Fault tree analysis is one of the tools of reliability engineering.[2] Other tools include event trees, failure mode and effect analysis and the concepts of probability, reliability and the (Boolean) algebra of 'AND' and 'OR' logic. Reliability engineering is a very important way of considering the reliability of different engineering options, for example, of different approaches to the design of interlock systems.[3] Reliability engineering and quantified risk assessment are, however, not the same thing. Quantification of risk for a major hazard chemical plant needs to consider the full range of accident causation and furthermore considers the acceptability of that risk. Quantified risk assessment may have grown from the traditions of reliability engineering but the two now need to be seen as separate and not synonymous.

Risk assessment from an engineering perspective is often considered to involve first an estimation of the risk and second an evaluation of the significance of the risk. This is illustrated in Figure 3.2. Risk assessment may involve comparison of the calculated risk (the risk which has been estimated) against risk criteria in the evaluation stage to judge the acceptability of the risk. However, the authors of Ref. 1 are careful to point out the limitations of Quantified Risk Assessment (QRA) and emphasise that acceptance of an activity should not be based on estimated risk levels alone. The authors say:

RISK ASSESSMENT	
Estimate the risk - How big is it?	Evaluate the risk - How significant is it? Is the risk acceptable?

Figure 3.2 *Risk assessment not only aims to estimate risk, but also to evaluate its significance, so as to consider if the risk is acceptable. These considerations are a crucial part of the human dimension to risk assessment. In risk assessment there is a human dimension in causing accidents and a human dimension in estimating risk and evaluating its significance. Thus the acceptance of risk is not straightforward*

The whole analytical exercise seeks to be objective. It must be realized, however, that as in many scientific and engineering exercises assumptions, estimates, judgements and opinions may be involved. Because of these limitations, formal quality systems are being increasingly introduced to document assumptions, employ recognized data sources and models, and thereby reduce variability. As with many techniques, they should be used by people who understand their limitations, and with caution.

3.1.2 Engineering Approaches – Some Further Issues

In considering engineering approaches to risk assessment a number of important issues and distinctions need to be raised.

Quantified risk assessment for a major hazard chemical plant can be performed at different levels of detail. A very detailed approach considers failures of items of hardware such as valves or pipes, and people errors across the whole of the plant and its life cycle, e.g. design, construction, operation and maintenance. The risk assessment builds up the failure events in terms of consequences and frequencies. This approach needs detailed knowledge of plant-item failure rates and human error rates, and implies complex modelling such as fault tree analysis. This detailed approach can be referred to as a 'bottom-up' approach because it 'builds up' the risk analysis from detailed information and detailed modelling. However, this may be contrasted with an 'overall' approach to risk assessment which can be taken by considering a reasonable set of failure possibilities – e.g. loss of containment from vessels, pipes and so forth – and assigning failure rates directly to these losses of containment possibilities. In this approach (a 'top-down' approach) it is usual to make use of generic failure rate data.

It is important to establish the conceptual differences between these 'top-down' and 'bottom-up' approaches. In the detailed 'bottom-up' approach individual items of hardware are considered. The failure rates used relate to hardware failure rates, while human error is modelled separately (see Section 3.2 on human reliability assessment). This approach most closely resembles the PRAs or PSAs (see Glossary) of the nuclear industry. In the more general, or top-down, approach it is release possibilities which are listed and directly assigned generic failure rates. This use of generic failure rates is crucial. This is because generic failure rates are taken to be representative 'averages' for the whole industry and so their use implies 'average' standards for safety culture, safety management, human competence and so forth. In addition, the use of generic data implies that all sources of failure have been included in the analysis and so the analysis can be considered complete.

The distinction between these detailed and general approaches is also important in other ways. The use of generic data implies an understanding only of the types of failure which can occur, e.g. a loss of containment of a toxic material from a pipe or vessel. The use of detailed modelling approaches implies that because the risk can be modelled the analyst understands all the causal mechanisms by which the failures occur. There are also very considerable differences in the amount of effort used in the two approaches. The detailed approach requires a large amount of effort while the 'overall' or general approach is more manageable. In the chemical industries the 'general' approach is much more commonly used. It is, therefore, a mistake to believe that 'QRA' is simply the chemical industry's name for 'PSA'. The two approaches may be very different in practice, although based on the same principles and they may have different implications for issues of data, resources and completeness in the analysis. In this context, completeness refers to a consideration of all issues relating to the causes of accidents, i.e. hardware, people and systems and cultures.

Risk assessment is a complex process and involves much judgement and uncertainty. The human dimension to risk assessment enters into the process of estimation and evaluation and, in addition, people may be the causes of accidents. Furthermore, risk as a concept is not straightforward and therefore it is clear that the process of comparing estimated risks with risk criteria has to be approached with care. There are at least five distinct reasons for caution which can be deduced from the discussion presented in this book.

- The risk assessment may be incomplete; important failure mechanisms may be under-represented or overlooked entirely.
- The subjective/objective debate shows that all risk assessment is at least partly subjective and will involve assumptions, estimates and judgements. Are these reasonable and transparent?
- The work of Fischhoff on acceptable risk shows that there are no universal values for acceptable risk. Acceptable risk is a decision problem. It depends on the risk to whom, when and in what circumstances.
- The risk discussion may be part of a wider debate and involve other interests which may be political or economic, for example.
- Those who benefit from the risk-creating activity and those who 'bear' the risk may be different people.

These issues are further developed later in this book.

An important issue of complexity and misunderstanding in risk assessment arises from the attempt to consider both the likelihood and the consequences of events. For example, in estimating individual risk of death for a series of hypothetical releases of a toxic chemical, the likelihood of each event is estimated and for each event this likelihood of the event occurring is used to estimate the likelihood of death given that the event occurs. For all the events these risks may be added to give an estimate of the individual risk of death. Confusion arises because this calculation involves a combination of frequency and consequence, but this is a combination of frequency for a series of events which all have the same outcome, not a combination of a series of different events with different outcomes.

Because of this confusion it is common practice to 'define' risk as the product of frequency and consequence. This is only true in the sense described above in which a series of events with different frequencies but the same outcome are combined. It is not true in the sense that a combination is made of events with different frequencies and different outcomes. As noted before it is not possible to compare, in a mathematically meaningful way, the risks from events with different outcomes.

3.2 HUMAN RELIABILITY ASSESSMENT

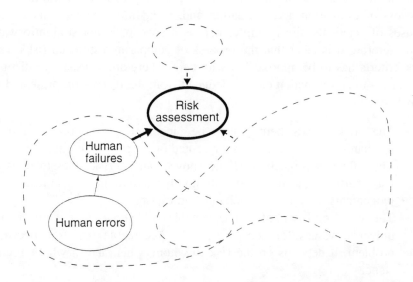

Figure 3.3 *Methods for human reliability assessments have been developed. These methods provide a quantified estimate of human error probabilities which can be factored into a risk assessment*

As mentioned above, in engineering approaches to risk assessment it is possible to attempt to quantify the contribution made by human action or inaction to the overall risk from a system. This quantification is mainly concerned with human actions or omissions as direct causes of accidents, not aspects of human failing which result in poor design or bad decisions. These approaches, known as Human Reliability Assessment (HRA), include the process of task analysis which helps with the identification of all points in a sequence of operations at which incorrect human action, or the failure to act, may lead to adverse consequences for plant and/or for people. HRA techniques assign a degree of probability on a numerical scale to each event and, by aggregating these, arrive at an overall figure for probability of human error. There is now a growing number of these methods which provide such quantification.[4-8] These tend to be described by acronyms such as THERP, HEART, SLIM and so forth. It is not my purpose to describe these various approaches in detail but it is important to emphasise that this area of work has received a considerable amount of effort over the years. The development of HRA received a strong impetus in the early 1980s following the accident at Three Mile Island (1979), which highlighted the role of the human operator and the catastrophic potential of human errors.

Thus, HRA approaches enable human error to be 'factored' into the risk assessment such that both engineering reliability and human error are considered together. However, it should be noted that the 'hardware' data will often include an element of human error, e.g. a valve reliability figure may include effects of how the valve was maintained, but the idea of HRA is to identify and to try to model points in a system where the human being can most easily degrade the system. Hence, HRA looks at individual human error, and its effect on system reliability.

HRA needs to be understood and used within an understanding of how human errors occur as described in Chapter 2 above. This point is clearly illustrated and discussed in the ACSNI publication on human reliability assessment.[4] The study group report that skill-based errors (slips) are amenable to prediction either from laboratory experiment, or from experience in other skilled tasks, even when those tasks are performed in different industries. The probability of a skilled typist striking the wrong key depends upon the nature and complexity of the material being keyed, but on average the error rate turns out to be much the same in an office of any kind as in laboratory experiments.

Rule-based errors on the other hand are more dependent on time, on individual cast of mind, and on the temporary physical, mental or emotional state of the human concerned. They are therefore harder to predict from past statistics or from experiment. Knowledge-based errors are even more difficult to predict. They arise from the detailed knowledge of the system possessed by the human and the resulting 'mental model' may be incorrect.

HRA is on much safer ground in predicting slips than in predicting mistakes. If two designs for a system are compared, and the error rate due to slips is found to be different, this is likely to be valid whatever the actual rates may be. It is therefore possible to compare the relative merits of systems, but there may also be some errors due to mistakes or violations that will make absolute predictions of overall error less reliable.

The ACSNI report continues, however, to say that caution about the possibility of quantifying mistakes should not obscure the value of quantifying slips; it is both desirable and valuable to reduce the area of uncertainty even if it cannot be eliminated altogether. Should an unacceptably high incidence of slips be predicted in a certain part of the system, steps should at once be taken to improve the situation without waiting for further advances to allow further and better prediction of mistakes. Both experiment and experience show that the probability of slips is rarely less than one in ten thousand, even in optimum conditions. It is essential therefore to take at least this level of error into account in assessing the overall safety of a system, rather than confining numerical estimates to the physical aspects alone.

There are therefore a number of cautions and qualifications that must be borne in mind when interpreting the results of HRA analyses. However, if a system involves humans, no complete risk assessment can be undertaken unless some method is used for introducing the possibility of human error into the overall estimate. The virtue of using HRA to do this is that it ensures structured and systematic thinking about each step in the human/machine chain of events. This means that points of particular hazard are less likely to escape attention. It also ensures that the results of such thinking are expressed in terms that allow comparison between different designs or parts of a design. Effort can then be concentrated upon improvement of key areas for safety.

HRA approaches try to model the effects of performance shaping (or influencing) factors on the probability of human error so that the effect of the immediate operating environment is accounted for. Thus, performance of a task is influenced by a wide range of factors. The person who performs the task can be considered as an individual, as a member of a team and as part of an organisation. In addition, the task is not performed in isolation but within a working environment. Thus performance shaping factors can include a very wide range of influences from the personal characteristics of the person, and the levels of stress and comfort under which the task is performed through to more complex issues such as levels of training and competency.

There are many useful accounts of HRA.[4-8] These describe similar strengths, limitations and weaknesses. It is important to emphasise that the fundamental point about these limitations and weaknesses is that our basic understanding of how human beings operate and think necessarily limits our ability to model the failure of people within systems. This may seem simple and obvious but we simply do not know how all the things people do (or do not do) can affect the reliability of engineering systems.

3.3 SAFETY MANAGEMENT STANDARDS AND QUANTIFIED RISK ASSESSMENT

3.3.1 Modification of Risk Approaches

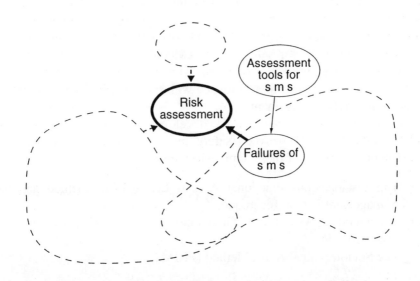

Figure 3.4 *Assessment tools for safety management systems (sms) have been developed. These audit tools provide a measure of the quality of the safety management system on a site. This quality will in turn affect accident causation and risk*

As mentioned above, risk assessment as used within the process industries often makes use of generic failure rates for loss of containment from items of plant such as pipework and vessels. In this context, generic failure rate data are derived from both historical and theoretical data and may contain elements of engineering judgement and caution. A single 'generic' value is taken to be representative of these data. As explained above, this is very important because the use of generic data is taken to mean that the risk estimate for the plant will represent 'average' values for the industry in terms of safety management standards and culture and implies that the assessment is complete. However, there is a growing awareness that these generic data may not be appropriate for use in all circumstances unless the role of management and organisational factors within the data is identified so that differences in safety standards for different plants which result in ranges of accident performance are made explicit. The use of generic data fails to reflect these differences. Another reason for using non-generic data is that industry is under strong pressure to improve safety management, but is unable to demonstrate the benefit of this, in terms of reduced levels of estimated risk, if generic data are still used within any risk assessment.

These concerns are well expressed by T. K. Jenssen in his paper *Systems for Good Management Practices in QRA.*[9] Thus he says, "In computing the risks, failure data from world-wide accident statistics are often being used. Little or no adjustment is being made for the quality of the safety management at the site being analysed. However, since most accidents are caused by the failure of management systems, QRA must take account of the local safety management practices, or else it will not provide reliable information for risk communication and for setting future priorities." He continues, "This calls for safety audits (which assess standards of Safety Management) to be used in association with the QRA, in order to correct the historical failure rates in accordance with the strengths and weaknesses of the local safety management."

Thus the past decade has seen the development of methods to describe and quantify the links between safety management and quantified risk assessment. Examples of these approaches are:

- The Instantaneous Fractional Annual Loss (IFAL) method and the Management Factor Technique.[10–12]
- Management Assessment Guidelines in the Evaluation of Risk (MANAGER).[13–16]
- The Sociotechnical Audit Method (PRIMA).[17–25]

These methods involve the estimation of risk using modified generic failure rate data to reflect standards of safety management. The modification of the risk is based on a site-specific audit which assesses the standards of safety management. If quantified risk assessment is used to estimate the risk using generic failure rate data, it is generally assumed that the risk estimate will be representative of at least an 'acceptable' or average plant with respect to standards of safety management. When the risk assessment is carried out using modified generic data the risk estimate is modified on the basis of the assessed standard of management on the site. This produces a risk estimate which is site-specific in the sense that it incorporates the perceived standards of management on the site.

The IFAL method with the management factor was the pioneering work in this field. However, for various reasons it has not been widely used, and experience with the system is not widely described in the literature.

The MANAGER system was also early pioneering work. It has been applied in a number of instances and experience with the system described.[13–16] Application of MANAGER has produced findings indicating that management influences could reduce risk estimates based on generic failure rate data by about half an order of magnitude (a factor of five) or increase them by about an order of magnitude (a factor of ten).

The sociotechnical audit method has involved extensive research and in-depth analysis of loss of containment from pipe-work, hoses and vessels, especially looking at underlying causes. This analysis is used to provide both a theoretical model to structure the audit question-set and a statistical basis to weight the audit areas.[17–25] Other workers have extended the use of this system looking especially at the issue of considering management influences at greater levels of detail within the risk assessment process.[26]

Separately, there are also developments which aim to include organisational factors in risk assessment which have been developed within the context of the nuclear industry.[27,28] One approach is the work process analysis model (WPAM). The goal is to provide a link between organisational factors, work processes and probabilistic safety assessment parameters, in order to facilitate the quantification of the impact of organisational factors on plant safety. This is achieved by calculating new (organisationally dependent) probabilities so that each new probability contains in it the effect of organisational factors.

These latter developments have taken place against the background of the state of the art in current PSA methodology which is such that organisational dependencies between hardware failures, between human errors, and between hardware failures and human errors are not modelled explicitly. Instead, the current methodology is confined mostly to models of isolated human errors and equipment failures.

A similar concept is described by Moieni and Davis.[29] Here a systematic approach is developed for incorporation of organisational factors into human reliability estimates to extend the applicability of probabilistic risk assessment to safety culture improvement and integrated risk management. Other systematic approaches have been described to model the influence of organisational factors on risk.[30–32]

It is clear that there are important links between the quality of safety management at a plant and any risk assessment carried out at the same site. The approaches described here have helped to establish these links in a quantitative manner. Standards of safety management are an important factor in determining risk and any assessment which fails to examine the effects of this variability will lack completeness. These approaches allow the assessor to ensure that, as one pivotal element of the assessment, key safety management systems are fully included within the assessment of risk for the site.

3.3.2 Accident Rates and Safety Management

When improvements are made to safety management systems on a site, it is natural to look for evidence that this improvement is affecting accident rates on the site. These accident rates are taken to be a measure of risk on the site. One approach to assessing the quality of safety management systems is to use safety management audit methods as described in Section 3.3.1. A widely used example of such an audit method is the International Safety Rating System (ISRS). When ISRS is used on a site, a measure of the quality of the safety management is represented by the percentage 'compliance' with the audit. Figure 3.5 illustrates this for one example taken from the literature.[33] Over the period 1984–1990 the percentage compliance with ISRS increased from 25% to 70% while the injury rate (risk) decreased from 5.2 to 0.88 (arbitrary units).

There are questions to be asked about the extent to which injury rates for individuals are a measure of plant safety with respect to the control of major hazards (notwithstanding difficulties in obtaining good injury-rate data). In Ref. 25, strong correlations were found between measures of safety management performance (which reflect major hazard safety) and personal involvement with accidents. The suggestion is that the same

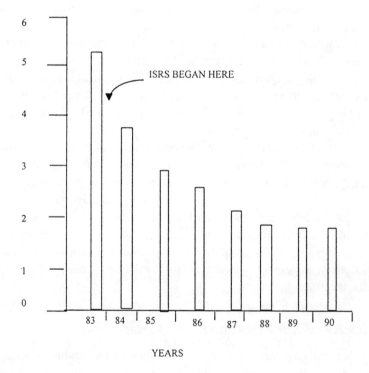

Figure 3.5 *Improvements to safety management can be measured by a reduction in accident rates on a site. Improvements to the safety management systems are also measured by audit methods such as the International Safety Rating System (ISRS). When ISRS is introduced the accident rates decrease on a year by year basis*
(Modified from Ref. 33)

The Assessment of Risk – Quantification

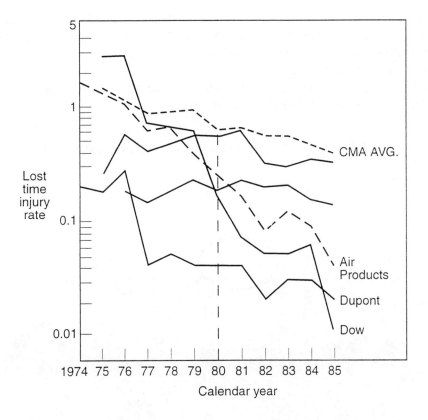

Figure 3.6 *Lost-time injury rates for different companies as a function of time. CMA AVG is the average for chemical manufacturers. For 1980 there is a range in injury rates from 0.7 for the average to 0.04 for the best performing company*
(Modified from Ref. 34)

causal mechanisms are at work, and that the risks to individuals of injury and the risks of a major chemical accident are affected by the same basic mechanisms and can be 'measured' by an audit of safety management systems. Furthermore, such reductions in accident rates show that risks on a site can be reduced by effective management while the technology of the process remains largely fixed.

A similar theme is shown in Figure 3.6 which is reproduced from Ref. 34. In Figure 3.6 lost-time injury rates for a number of companies are shown as a function of time. The figure shows two main points. Firstly, that accident rates (risks) can be reduced by a programme of safety management improvement and secondly that there is a range of performance in terms of accident rates (risks) for companies doing similar work. For example, in 1980 the chemical manufacturers' average accident rate (CMA AVG) is 0.7 (arbitrary units) while the 'Dupont' figure is 0.04. This is a ratio of 17.5 between 'very good' and 'average', i.e. risks to two individuals on two sites may vary by a factor of 17.5 based on this information even though they are involved in work of a similar general type. The figure only shows companies which are better than average. Assuming a similar range of 17.5 between the 'average' and the 'very worst' implies a range of 300 in accident rate variation between the 'very worst' and the 'very best' chemical manufacturers.

3.4 SAFETY CULTURE AND QUANTIFIED RISK ASSESSMENT

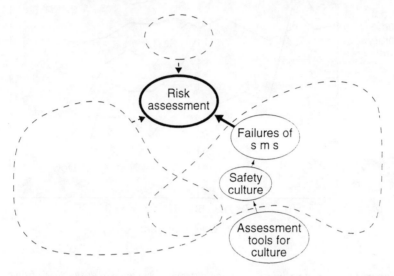

Figure 3.7 *Assessment tools for safety culture have been developed. These tools provide a measure of the strength of the safety culture which in turn affects accident causation and risk*

Chapter 2 reviewed some of the work which has pointed to the connection between a strong safety culture and low accident rates, where low accident rates indicate low risks on the site. In particular, the third ACSNI report,[35] based on a wide variety of evidence, concluded that different organisations doing similar work are known to have different safety records, i.e. good safety culture helps to ensure the safe operation of the plant. Such issues as adequate resources, good communications, co-operation which ensures the balancing of pressures which may arise from health, safety and production needs are highlighted and reconciled. Conversely, a poor safety culture will lead to high accident rates, i.e. the risks will be higher.

To try to quantify the nature of this relationship, information about the underlying attitudes and beliefs of the safety culture can be collected via surveys of attitudes. The results of such an attitude survey can then be used such that robust scales of attitudes are constructed and these are correlated with accident performance data for the companies concerned.[36–38]

For example, Refs. 36 and 37 contain information about a safety attitude questionnaire developed by the Safety Research Unit now at the University of Liverpool, which was used to measure attitudes to safety at major hazard plants. In this case the safety attitude questionnaire was distributed to participants and returned directly to the researchers. This ensured the confidentiality of the responses. Participants were asked to answer a set of questions such as "To what extent are your workmates satisfied with the safety procedures they are required to follow?". The respondents give a rating on a score which varies between 1 and 7 corresponding to very strong agreement or to very strong disagreement. The results of the safety attitude surveys are presented in the form of scores on attitude scales, which represent the combined responses to a number of questions which address the same issue. Examples of attitude scales are 'workforce satisfaction', 'safety information' and 'safe working procedures'.

Attitude scales for a particular site indicate a plant's relative score, on each of the scales, compared with a database of companies. In addition, self-reported accident rates (the percentage of respondents who report having been involved in an accident of any kind in the 12 months previous to the survey) are reported in the safety attitude questionnaire, and strong correlations were found between the self-reported accident rates and the attitude scale scores.

Another approach in which accident rates are specially monitored as a way of demonstrating improved safety is the behaviour-based approach to safety.[39-41] In this approach, observable and measurable actions critical to safety at a site are monitored. Such actions are typically points of contact on a ladder, stairs or catwalk, body position in relation to a task and wearing protective equipment. This approach seems to be most relevant to occupational safety rather than major hazard risks.

3.5 SUMMARY AND DISCUSSION

This chapter has built on Chapter 1 and Chapter 2. From the perspectives of safety engineering, human errors and systems and cultures the links with quantified risk estimation have been briefly described.

Engineering approaches to risk assessment have been widely developed and applied. Such risk assessments may involve comparison of the estimated risk with risk criteria to judge the acceptability of risk. However, risk assessment is a complex process and involves much uncertainty. Furthermore, risk as a concept is not straightforward and therefore it is clear that this process of comparing risks with risk criteria has to be approached with considerable care. Five distinct areas for caution are suggested:

- The risk assessment may be incomplete. Important failure mechanisms may be under-represented or overlooked entirely.
- The subjective/objective debate shows that all risk assessment is at least partly subjective and will involve assumptions, estimates and

judgements. Are these reasonable and transparent?
- The work of Fischhoff on acceptable risk shows that there are no universal values for acceptable risk. Acceptable risk is a decision problem. It depends on risk to whom, when and in what circumstances.
- The risk discussion may be part of a wider debate and involve other issues and interests, which may be political or economic, for example.
- Those who benefit from the risk-creating activity and those who bear the risk may be different people.

Within engineering accounts of risk assessment, the limitations of risk assessment are generally accepted and the simplistic use of risk criteria is guarded against.

The important role of human error in risk assessment has led to a range of methods which attempt to quantify the role of human error as a contributory factor to the overall risk. This has happened against the background of the considerable work done to understand human error which was briefly described in Chapter 2. There are considerable advantages in attempting to include the role of direct human errors in risk assessment, but it is important to emphasise that fundamental limits to our understanding of how human beings operate and think necessarily limits our ability to model the failure of people within systems. Nevertheless, human error modelling in risk assessment is now well established and is often included within the engineering approaches.

Attempts to view accident causation and risk assessment in a holistic manner have led to the development of methods which consider the overall system and try to describe and quantify the links between safety management, safety cultures and quantified risk assessment. These approaches may use generic failure rate data modified to include perceived standards of safety management in an attempt to avoid the problems of lack of complete understanding of how the system might fail, but nevertheless to reflect in the risk estimate perceived standards of safety management and safety culture. These approaches are not widely used and may be classified as 'experimental'. Furthermore, the precise nature of the relationship between the assessment audit tools and the theoretical description of systems and cultures (Sections 2.1.3–2.1.7) is not transparent.

The subjective/objective nature of risk was introduced in Chapter 2. In extending this concept into the area of risk assessment it is important to emphasise that risk assessment is neither subjective nor objective absolutely. Risk assessment will always be subjective, at best to a small degree but possibly to a large extent. This is illustrated in Figure 3.8. On the objective end of the scale (the left-hand side) risk estimates make use of data from long periods of experience or consider simple compari-

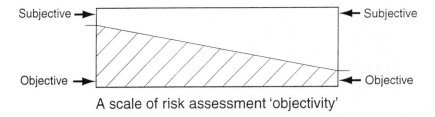

Figure 3.8 *A scale of risk assessment objectivity. Some risk estimates are more objective than others. This can be referred to as 'relative objectivity' to reflect the extent to which data, information and other evidence enter into the estimation process*

sons 'before' and 'after' safety improvements. Thus accident statistics for, say, road use are well established and an estimate of the risk of death from car driving might be called objective. Nevertheless, this involves judgement and is still only related to 'average' figures for certain classes of driver covering specified mileage in particular vehicles.

Another example of 'objectivity' in describing risk is provided by the Health and Safety Executive's work on a certain type of train door. In the past these doors sometimes came open when the train was moving and people fell out. This had been going on for years and about 20 people a year were being killed. It was established that the doors could look as if they were properly closed when in fact the bolt was jammed inside the lock and not engaged at all. The number of people killed after improvements to the locks were made was down to three per year. It would seem to be 'objective' to consider that the risk of death from falling from such a door had been reduced by this intervention. Furthermore, it would be reasonable to quantify this reduction in risk.

On the subjective end of the scale (the right-hand end) of Figure 3.8 the risk assessment may relate to an estimate of risk for an accident which has never occurred at a chemical plant which has not been built. The risk estimate which considers the probability of death of a hypothetical individual living near to such a plant will necessarily be to a significant extent subjective. Inevitably, the judgements of the risk assessor will be important. Nevertheless, objectivity enters the assessment and at the most basic level consideration of the types of chemicals, their quantities and toxicities are important objective issues to be considered in the assessment of the risk. It is also relevant to consider the historical experience of similar plants.

Other authors have described risk in similar terms. Thus the Royal Society Study Group[42] (see Chapter 2) describe:

- Risks for which statistics of identified casualties are available.
- Risks for which there may be some evidence but where the connection between suspected cause and injury to any one individual cannot be definitely traced.
- Risk estimates for events that have not yet happened.
- Risks that are not foreseen.

These four classes would seem to represent different positions on a scale of objectivity for risk assessment. We might wish to use the term 'relative objectivity' to describe estimates of risk in terms of the extent to which evidence enters into the assessment process.

Similarly Otway[43] says "There is a great deal of difference in the certainty that can be attached to the size of different risks. National statistics on deaths due to common accidents show only minor fluctuations from year to year, but for newer technologies the only way to quantify their risks is by theoretical calculations because there has not been enough experience to form a statistical basis."

So risk assessments may vary in their objectivity and completeness. It may appear, that by increasing the completeness of the risk analysis, the process has apparently moved away from the desired objectivity of science and mathematics where engineering approaches to risk assessment have their roots. However, this change is more apparent than real because engineering approaches to risk assessment necessarily involve a considerable element of judgement and uncertainty (see Section 3.1).

In considering the accident performance of different chemical sites it has been shown in Chapters 2 and 3 that there is a range of actual risk performance when different organisations carry out similar work. This is due to both organisational factors and contextual factors, i.e. to factors which affect the way a plant is organised and run and to the environment in which it operates. Essentially there is a scale of performance with good performers (low risk organisations) at one end and bad performers (high risk organisations) at the other. The argument was introduced in Chapter 2 and further developed in Chapter 3 that this range of performance represents differences in safety management, safety culture and organisational structures.

In Figure 3.9 the idea of a scale of risk or accident performance is illustrated in relation to the theoretical perspectives which describe organisational structure, safety management and safety culture (Sections 2.1.3–2.1.7). Thus Figure 3.9 builds on Figure 2.6 by introducing a risk scale which shows low accident rates for organisations at the 'ideal' end of the scale and higher accident rates for organisations which show an increasing 'distance' from the ideal.

However, the role of engineering should be very clearly stated in this scale too. Of course, good engineering and a sound understanding of

The Assessment of Risk – Quantification

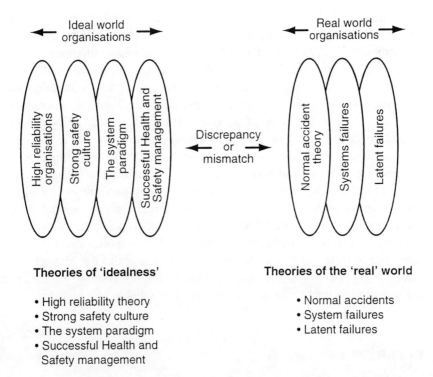

Theories of 'idealness'

- High reliability theory
- Strong safety culture
- The system paradigm
- Successful Health and Safety management

Theories of the 'real' world

- Normal accidents
- System failures
- Latent failures

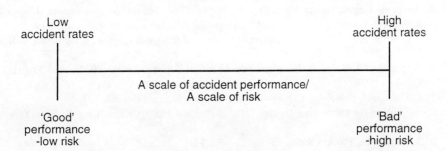

Figure 3.9 *A scale of accident performance (a scale of risk) related to organisational characteristics. Ideal organisations are described by theories of idealness. These organisations have low accident rates (low risk). Real world organisations show a significant difference from the ideal. They have higher accident rates and higher risks. The greater the difference between the ideal and the real world the greater the risk difference, provided technological considerations are equivalent*

scientific principles as a source of reliability have been historically at the forefront because of the enormous advances which have been made in safety through the applications of good engineering principles and the occupational hygiene principles of health. An illustrative example here is the effects of good ventilation systems in mines. Understanding the phenomenon of lower flammable limits of methane/air mixtures (methane/air mixtures with less than 5% methane are not explosive) and avoiding mixtures above the lower flammable limit by appropriate ventilation makes the difference between large scale accident potential and safe operations. In more recent years the scale of good and bad performance (Figure 3.9) has tended to concentrate on issues of safety culture and of organisational structure and management but against an *established* background of good engineering controls and good scientific understanding of accident causes. It would seem to be valid to compare the accident performance of different sites in terms of differences in safety management, organisational structure and safety culture provided that technical considerations are equivalent, e.g. to compare two sites performing essentially similar work or to compare the same site at different times before and after changes to the management of safety. Nevertheless, the features of, for example, a strong safety culture are expected to lead to relatively low accident rates in any industry.

REFERENCES

1. R. Pitbaldo and R. Turney, *Risk Assessment In the Process Industries*, Institution of Chemical Engineering, Rugby, 1996.
2. B. Skelton, *Process Safety Analysis – An Introduction*, Institution of Chemical Engineering, Rugby, 1996
3. R. A. Freeman, *Process Safety Progress*, 1994, **13**, 146.
4. Advisory Committee on the Safety of Nuclear Installations, *Second Report: Human Reliability Assessment – A Critical Overview*, HMSO, London, 1991.
5. I. Glendon and E.F. McKenna, *Human Safety and Risk Management*, Chapman & Hall, London, 1995.
6. B. Kirwan, *A Guide to Practical Human Reliability Assessment*, Taylor and Francis, London, 1994.
7. E. Hollnagel, *Reliability Engineering and Systems Safety*, 1996, **52**, 327.
8. Guidelines for Preventing Human Error in Process Ind. CCPS. AIChemE, New York, 1994.
9. T. K. Jenssen, Process Safety Progress, 1993, **12**, 137.
10. J. Powell and D. Canter. J. of Environmental Psychology, 1985, **5**, 37.
11. J. Singh and G. Mundy, 'A Model for the Evaluation of Chemical Process Losses', Design 79 Symposium, paper c-1, Institute of Chemical Engineering Midlands Branch, Rugby, September 1979.
12. D. Canter and J. Powell, in *Human Decision Making and Manual Control* Ed. H.P. Wilhumeit, Elsevier Science Publishers (North-Holland), Amsterdam, 1986, p. xx.
13. L. J. Bellamy and T. A. W. Geyer, 'Techniques for assessing the effectiveness of management. Paper presented to European Safety and Reliability Research Development Association (ESReDA) Seminar on Human Factors, Bournemouth, March 1988.
14. Technica Ltd. *The MANAGER Technique*, Technica Ltd., London, 1988.
15. R. M. Pitblado, J. C. Williams and D. H. Slater. *Plant Operators Progress*, 1990, **9**, 169.

16 J. C. Williams and N. W. Hurst, A Comparative Study of the Management Effectiveness of Two Technically Similar Major Hazard Sites. In Major Hazard Onshore. 20–22 Oct. 1992 UMIST. Manchester. I Chem E symposium Series 130 Page 73. 1992.
17 N. W. Hurst, L. J. Bellamy, T. A. W. Geyer and J. A. Astley, J. Haz. Mat. 1991, **26**, 159.
18 N. W. Hurst, L. J. Bellamy and T. A. W. Geyer, Organisational, Management and Human Factors in Quantified Risk Assessment. A theoretical and Empirical Basis for Modification of Risk Estimates. Safety and Reliability in the 90's (SARRS '90). Ed. Water and Cox. Elsevier Applied Science, London and New York, 1990.
19 N. W. Hurst, *Immediate and Underlying causes of Vessel Failures: Implications for Including Management and Organisational Factors in Quantified Risk Assessment*, Symposium Series No. 124, Institute of Chemical Engineering, London, 1988, p. 155.
20 N.W. Hurst, R. Hankin, J. Wilkinson, C. Nussey and J. Williams, 7th International Symposium on Loss Prevention and Safety Promotion in the Process Industries, Italy, 4–8 May, 1992
21 N. W. Hurst, L. J. Bellamy and M. S. Wright. Research Models of Safety Management of Onshore Major Hazards and their Possible Application to Offshore Safety, Symposium Series No. 130, Institute of Chemical Engineering, London, 1992, p. 129.
22 N.W. Hurst, R. Hankin, L. Bellamy and M. Wright, J. Loss Prev. Process Ind., 1994, **7**, 197.
23 N.W. Hurst, J. Davies, R. Hankin and G. Simpson, 'Failure Rates for Pipework – Underlying Causes', Paper presented to Valve and Pipeline Reliability Seminar, University of Manchester, Institute of Mechanical Engineering, February, 1994.
24 N. W. Hurst, Auditing and safety management CECDGXII/ESReDA Conference "Operational Safety Seminar" Lyon France 14–15 October 1993.
25 N.W. Hurst, S. Young, I. Donald, H. Gibson and A. Muyselaar, J. Loss Prev. Process Ind., 1996, **9**, 161.
26 I.A. Papazoglou and O. Aneziris, Quantifying the Effects of Organisational and Management Factors in Chemical Installations, in, C. Cacciabue and I. Papazoglou, eds., Probabilistic Safety Assessment and Management, 1996, ESREL '96, PSAMIII, Vol. 2, pp. 922–927.
27 K. Davoudian, J. S. Wu, and G. Apostolakis, *Reliability Engineering and System Safety*, 1994, **45**, 85.
28 K. Davoudian, J. S. Wu, and G. Apostolakis, *Reliability Engineering and System Safety*, 1994, **45**, 107.
29 P. Moieni and D. Davis, *An Approach for Incorporation of Organisational Factors into Human Reliability Analysis in PRAs*, Accident Prevention Group, San Diego, 16980 via Tazon, ste. 110, CA 92127.
30 D. E. Embrey, Reliability Engineering and System Safety, 1992, **38**, 199.
31 M. Modarres, A. Mosleh and J. Wreathall, Reliability Engineering and System Safety, 1992, **38**, 157.
32 M. E. Paté-Cornell and D. M. Murphy. Reliability Engineering and System Safety, 1996, **53**, 1.
33 B. Arnold, 'Safety – A catalyst to improved organisational performance', 7th International Symposium on Loss Prevention and Safety Promotion in the Process Industries, Italy, 4–8 May 1992.
34 P. L. Thibaut Brian, *Managing safety in the Chemical Industry*, Symposium Series No. 110, Institute of Chemical Engineering, London, 1988, 615.
35 Advisory Committee on the Safety of Nuclear Installations, *Third Report: Organising for Safety*, HMSO, London, 1993.
36 N.W. Hurst, S. Young, I. Donald, H. Gibson and A. Muyselaar, J. Loss Prev. Process Ind., 1996, **9**, 161.
37 I. Donald and D. Canter, J. Loss Prev. Process Ind., 1994, **7**, 203.

38 S. Cox and T. Cox, Work and Stress, 1991, **5**, 93.
39 J. H. Hidley and T. R. Krause, *Professional Safety*, October 1994, p. 28.
40 T. R. Krause, The Behaviour-based Approach to Safety, The Safety and Health Practitioner, August 1991, p. 30.
41 L. Finlayson, T. Fishwick and A. Morton, Loss Prev. Bull, **130**, 3.
42 Royal Society, *Risk: Analysis, Perception and Management*, Report of a Royal Society Study Group, Royal Society, London, 1992.
43 H. Otway, *Regulating Industrial Risk*, Butterworth, London, 1985.

CHAPTER 4

Risk and Decision Making

Chapter 1 of this book examined a number of case studies which illustrate the causes of accidents from the three perspectives of safety engineering failures, human error and the failure of systems and cultures. The pursuit of 'completeness' in describing accident causation was considered to be important because it provides the best opportunity for learning for the future from the mistakes of the past. In this context, therefore, 'completeness' refers to a full consideration of issues relating to hardware, people and systems and cultures.

In Chapter 2 these different perspectives of accident causation were further developed by a consideration of the theoretical approaches which essentially build on these different approaches for explaining accident causation. It was apparent that health, safety and loss prevention had been extensively developed and applied within the chemical engineering discipline. In addition there has been extensive work done to understand human error in a wide range of situations which include hazardous technologies. More broad-based accounts of organisations and their functions were described under the heading of systems and cultures. Considerable general commonality was found between some aspects of these different descriptions. This finding of commonality and overlap between the different approaches to describing 'ideal' organisations and the different approaches to describing 'real world' organisations was an important finding in Chapter 2. Thus, one observer may emphasise the importance of safety culture and another safety management, but in practice there is overlap between these descriptions.

Chapter 3 introduced the concepts of quantification of risk (risk estimation) and looked at how risk may be quantified in relation to the perspectives of safety engineering, human error and systems and cultures.

It was also shown in Chapters 2 and 3 that there is a range of risk performance when different organisations carry out similar work. Essentially there is a scale of accident performance with 'good performers' (low risk companies) at one end and 'bad performers' (high risk companies) at the other. The argument was introduced that this range of risk performance may under appropriate circumstances represent differences in safety management, safety culture and organisational structures. Chapter 3 (Figure

3.9) made a link between the descriptions of 'ideal' and 'real world' organisations and the idea of a range of performance, i.e. high risk and low risk companies.

It has also been shown that risk is a complex concept, and so the use of estimated risk values as an aid to decision making concerning the acceptability of risk is complicated and must be approached with caution. This complexity does not relate to the inability of some people to understand what is clear to others but rather to the difficulty of understanding interlinked and interdisciplinary concepts and genuine differences in need and perspective. Chapter 3 suggested five distinct reasons for caution in acceptable risk decisions (Section 3.1.1).

In addition, a confusion has arisen that risk is a product of probability and consequence. This is only strictly true when a series of events with different frequencies but the same outcome are combined. It is not true when a combination is made of events with different frequencies and different outcomes.

In Chapter 4 two of these themes will be further explored. Examples of risk-based decisions will be introduced and discussed to consider the complexity of risk as a concept and to illustrate the caution which needs to be taken when risk estimates are used in decision making, and the need to consider the full range of human dimensions in risk assessment.

The overlap and commonalities which have been shown to exist between different approaches and descriptions to 'ideal' and 'real world' organisations are further developed by considering how, when these differences in 'idealness' between two companies, or sites, are measured, the measurement might be related to the scale of risk performance which was introduced in Chapter 3 (Figure 3.9). This is discussed within the context of completeness in risk assessment processes.

4.1 RISK-BASED DECISIONS

4.1.1 Is Risk a Useful Concept?

The complexity of risk as a concept has been illustrated in this book. Moreover, there would seem to be confusion over quantified descriptions of risk, which only makes the situation worse. The concept of hazard, on the other hand, seems to cause less of a problem, and is variously described as 'objective' or 'real' as opposed to risk which is described as 'subjective' and 'constructed'. Against this background it seems reasonable to ask the question "Why estimate risk at all?" What are the attractions of using estimated risk values as an aid to decision making? The answer to this question seems to come from a desire to make decisions which are rationally based. The use of estimated risk values is considered to hold the promise of rational decision making in a range of circumstances:

- In making investment decisions to improve health and safety, the aim would be to maximise the risk reduction achieved against the investment.
- Risk assessments have potential to compare different options for achieving the same desired improvement to health and safety and again maximising the risk reduction for the cost incurred.
- Investment to remove a hazard may not seem sensible if the likelihood of realising the hazard is judged to be very low, i.e. the risk to any individual is low.
- Consideration of high consequence, low likelihood events.

Risk assessment has been widely applied in considering incidents with very low likelihood but very high consequence, e.g. the escape of all the toxic gas from a large chemical store. Thus, for a risk assessment for a chemical plant it would be usual to consider all potential events leading to a loss of containment of the toxic material, especially when risk to the public outside the site boundary is of interest. The risk assessment would often include a consideration of the hypothetical complete failure of any storage vessel on the site, but because the likelihood of this event is very low this event does not dominate the estimated risk to any one individual outside the boundary. However, when such an event is considered only in terms of the hazard, e.g. by considering the geographical extent of the harm caused by the realisation of the hazard, it will be shown to have a very large hazard range. It might typically be several miles from the potential release before all significant toxic effects on people are lost, but on the other hand the event may be judged to have a likelihood of once in a million years.

What this means is that from the perspective of hazard ranges, the greater the potential release volume the greater the distance will be before all significant toxic effects on people are lost, but from the perspective of risk (which includes the likelihood of the event) more minor events such as holes or leaks in vessels or pipes may be more significant in estimating the risk because they are more likely to occur. If the purpose of considering the chemical risk for the plant is to answer a question such as "How far from the plant is it safe for people to live?" then a consideration of hazard range will produce an answer of several miles, while a consideration of risk will produce an answer which is much closer to the chemical plant (see Section 4.1.4). Thus although the concept of risk is often considered 'subjective' and 'constructed' while hazards are 'objective' and 'real' the consideration of hazard ranges also has serious problems when used as an approach to this decision-making process. This is because it is not reasonable to consider large tracks of land unusable (or unsafe) on the basis of events which have a very high consequence but which are considered to be very unlikely to occur. Similar arguments apply to flood protection, or the strength of dams. The point is that the exclusive use of hazard potential does not produce a simple decision-making mechanism. As such, risk assessment appears to be, intrinsically, a useful approach to these types of problems, combining as it does both likelihood and outcome.

So risk estimates, as an aid to decision making, have the strength of considering both the extent of the hazard and the likelihood of its realisation and so hold the promise of investment in health and safety which is of maximum benefit. Thus risk assessment is often carried out as an aid to decision making, and these decisions may relate to such issues as:

- the acceptability of risk in particular circumstances;
- the priorities to be attached to certain risk reduction measures;
- consideration of different design options for achieving risk reduction.

4.1.2 Risk Assessment and Risk Ranking

Priorities to be attached to certain risk reduction measures on a particular site may be developed by risk-ranking approaches, but these need to be distinguished from the quantified risk assessment processes of the type discussed in this book and its references. Risk-ranking approaches attach a 'numerical' scale to two dimensions of 'risk' in terms of likelihood and severity. These numerical scales are no more accurate than using words. They are for illustration only and cannot be interpreted in any absolute or mathematical sense. Thus the scale of severity might be, for example, 1 (minor), 2 (serious) and 3 (major). Similarly, the scale of likelihood might be 1 (unlikely), 2 (likely) and 3 (very likely). These scales can be used to rank 'risks' on a particular site from 1 (a minor outcome which is unlikely) through to 9 (a major outcome which is very likely). Such risk ranking is often used to provide an overall view of a wide range of workplace hazards on a site, for example electrical, mechanical and chemical hazards. The aim would be to eliminate 'high risks', i.e. those judged to be both major or serious and very likely. The aim would also be not to spend resources on risks which are 'small' (minor and unlikely), while higher risks remain. The problem with such approaches is that they do not satisfactorily discriminate between the 'middle range' risks which may be, say, minor and frequent, or serious and unlikely, for example. Judgement is then needed to attach priorities to any risk reduction measures. In those circumstances other important factors are likely to be involved in any decision concerning which risks to reduce. These factors will often relate to the likelihood that the risk reduction measure will succeed and also to the cost of the proposed measure. Thus for two hazards which are judged to be of similar 'risk' it would be appropriate to prioritise a risk reduction measure which was judged to be very likely to reduce the risk, against a risk reduction measure which had a judged outcome that it might not reduce the risk so successfully. Similarly, for two risk reduction measures which were judged to be equally likely to succeed, then it would be reasonable to prioritise the measure with the lower cost.

Other authors have expressed similar reservations about risk-ranking approaches.[1,2] Some of the difficulties of such approaches are considered to include:

- the inability to distinguish between high frequency, low consequence events and low frequency, high consequence events;
- the fact that most risks fall in the middle of the range of risk;
- the challenge of priority setting, which is not only of ranking risks, but also of ranking solutions to risk problems.

Risk-ranking methods are initially attractive, combining likelihood and severity into a single number. But in practice the very low risks are often obviously low and the very high risks are also obvious (hopefully). The criticisms of not being able to rank the middle ranges of risk, and of the arbitrary nature of the ranking methods, mean these methods are not as useful as at first they appear.

The context of this book is risk assessment as applied to chemical and process industries. This is mainly concerned with risk in the sense defined in the glossary, i.e. likelihood of a specified outcome. The risk-ranking methods are not really applicable within this context but they are important because of the utility they have for ranking a whole range of hazards in some order of 'risk'. However, risk-ranking approaches have also caused confusion because of the proposition that risk = probability × consequence. So risk-ranking approaches do not estimate risk in the same way as quantified risk assessments. In fact, risk-ranking methods use arbitrary scales which are multiplied together in an arbitrary way to provide a 'numerical' estimate which is useful if considered with care, but which is not capable of strict interpretation.

Chapters 2 and 3 describe a range of risk performance when different organisations carry out similar work. This range relates to the likelihood of a specified outcome in order to be mathematically meaningful, not to risk ranking in the sense of an attempt to a rank a wide range of different risks on one site.

Essentially, risk-ranking approaches are applied within a single site to rank a wide range of hazards on the site in some order of 'risk', while different risk performance between sites would be represented by, for example, differences in injury rates between the sites, when a comparison is made of the likelihood of a specified outcome occurring between two, or more different sites.

Thus it would be meaningful to compare accident rates between two sites in terms of numbers of reported injuries of a specified severity, or numbers of chemical leaks of a specified size and duration, and provided some way is found of comparing the 'size' of the sites in terms of numbers employed or tonnage of product produced then the different risks on the two sites may be compared.

4.1.3 Risk/Cost Trade-offs

A separate point to consider in risk-based decision making is the assumption that there is a cost/risk reduction trade-off, i.e. that time and money have to be spent to reduce risk and that this erodes profit. It is useful to distinguish two situations:

- An immediate need to spend money to purchase equipment, e.g. personal protective equipment, or improve a design feature of a plant to reduce risk.
- Longer term issues concerning overall company viability and profit.

Clearly, expenditure on new safety equipment costs money and this expenditure may be seen to reduce the profits of the organisation. But this is a complex issue and careful studies show[3] that the overall costs of accidents are often largely in excess of the immediate costs. In addition, many organisations now consider accidents in their workplaces to be unacceptable.[4] This is because of the human cost of accidents and due to their desire to run an organisation which is considered a market leader, commands respect within the community and attracts the most able employees. Thus the notion of a cost/risk reduction trade-off is simplistic and not the only aspect of risk reduction which would in practice be considered in making decisions about risk reduction expenditure.

4.1.4 Land-use Planning Around Major Hazard Sites in the UK

The role of acceptable risk in decision making may be illustrated in relation to land-use planning around major hazard sites in the UK.[5,6] This relates to one of the long-standing tasks of the Health and Safety Executive in the UK; to advise local planning authorities on the development of land in the vicinity of existing major hazard installations.

The advice is based on estimated risk values around the site, which are obtained from a quantified risk assessment process called RISKAT[7,8] which makes use of generic data. Risk estimates based on this procedure relate to individual risk of receiving a specified 'dangerous dose' of toxic chemical (a combination of concentration and exposure time), not an individual risk of death. Once the risk estimation process is complete, the Health and Safety Executive will usually advise the local planning authority against developments of land near major hazards in the following circumstances:

- Housing developments providing for more than about 25 people where the estimated individual risk of receiving a defined 'dangerous dose' of toxic substance, exceeds ten in a million per year.
- Housing developments providing for more than about 75 people where the calculated individual risk of receiving a 'dangerous dose'

exceeds 1 in a million per year. The 1 in a million per year criterion may be extended downwards somewhat for developments in which the inhabitants are unusually vulnerable or the development is extremely large.

These estimated risk levels are often plotted as contours on a map of the location. An example is shown in Figure 4.1. Thus the advice which HSE gives about whether the risk is acceptable on health and safety grounds is based on whether the proposed developments fall 'outside' or 'inside' specified contours of risk. The specified contour is in turn dependent on the nature of the proposed development.

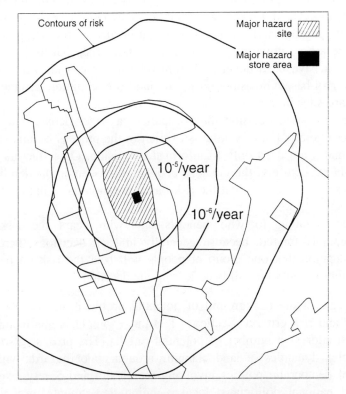

Figure 4.1 *Estimated individual risk contours around a major hazard site. Housing developments for more than 25 people would be advised against if the development fell within the 10^{-5}/year contour while a development for more than 75 people would be advised against if it fell within the 10^{-6}/year contour. 10^{-5}/year is equivalent to a frequency of 10 in a million years while 10^{-6}/year is equivalent to a frequency of 1 in a million years*

Recall from Chapter 3 (Section 3.1.2) that five distinct reasons were suggested for exercising caution in such acceptable risk decision making:

issues of completeness; the subjective element in risk assessments; the absence of universal values for acceptable risk; the fact that risk is often part of a wider debate; issues relating to who benefits and who bears the risk. In considering this particular acceptable risk-based decision problem the following observations can be made in relation to these five areas of caution:

- The process makes use of generic failure rate data within the risk assessment which helps to ensure that the data are an industry average and are representative and complete in the sense that major causes of past accidents will be included within the generic data.
- The methodology is applied in a consistent manner using standard data sources, models and calculation methods. This helps to overcome the subjectivity which is inherent in such a process.
- There are no universal values for acceptable risk, but in this case the criteria have been developed from relevant case studies and for this one risk-based decision only, i.e. land-use planning around major hazard sites in the UK.
- The risk-based decision may be part of a wider debate and involve other interests but this point is explicit in the process because the risk estimates are used only to inform the decision. Other factors can be made explicit, e.g. the creation of jobs, improvement to derelict land and so on. It is clear that the risk-based decision has a highly political aspect.[9,10]
- Issues relating to who benefits and who bears the risk seem acceptably treated, because generally in such decisions there is no implication that one group of people will bear the risk on behalf of another group who will benefit from the decision.

Thus, against the background of potential problems in the use of risk estimates and risk criteria (Section 3.1.2) this example seems to minimise or at least address a number of potential pitfalls. The process also shows the complex dynamics of land planning around major hazards within the context of democratic politics involving the Health and Safety Executive, local and national politicians, local planning departments, organisations involved with the planning application and local residents, i.e. it shows the inherently political nature of these decisions.

4.1.5 Storage of Nuclear Waste

In contrast to the development and use of land around existing major hazard sites in the UK, the decisions regarding the disposal of nuclear waste in the USA have been extremely problematical.[11,12] In 1962, for example, the risks associated with a proposed site for shallow land burial of radioactive wastes were estimated. Scientists from industry and academia praised the location and calculated that, if plutonium were buried there, it would take 24,000 years to migrate one-half inch. They said that the possibility of subsurface migration off site was non-existent. Yet only ten years after the facility was opened, plutonium and other radionuclides were discovered two miles off site. The predictions were wrong by six orders of magnitude (a factor of a million).[11]

Examples of this kind call into question the completeness of the risk estimation process, and therefore the extent to which the estimate is subjective and not objective. Furthermore, the decision on the acceptability of the risk is clearly part of a much wider debate which involves social, economic and political issues. These problems are exacerbated because the people expected to bear the risk do not benefit from the decision and only expect their locality to be stigmatised by the presence of nuclear waste. Overwhelmingly there is local opposition to such a proposal.

Clearly, consensus in risk-based decisions can be very difficult to arrive at and decisions difficult to implement if the full range of human dimensions in risk assessment are not fully addressed.

4.2 MEASURING RISK PERFORMANCE BETWEEN SITES – ISSUES OF COMPLETENESS

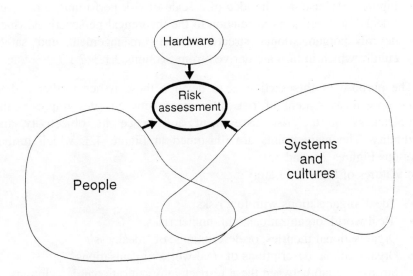

Figure 4.2 *Is the risk assessment complete? Are issues related to people and systems and cultures factored into the assessment? Are the assumptions clear and transparent?*

Thus, one important issue in acceptable risk decisions is related to the completeness of the risk assessment process.[13,14] This is because the process will be of itself more robust and therefore acceptable if it can be demonstrated to be complete. Conversely, if the process is shown to be incomplete then it is open to the charge of subjectivity. Incompleteness in risk assessment, therefore, has the potential to undermine its entire credibility and usefulness as an aid to decision making (see Section 4.1.5). This is illustrated in Figure 4.2.

In Section 4.1.4 the use of land around major hazard sites in the UK was discussed. It was suggested that the use of generic failure rate data was a strength in this process because it implied that the data are an industry average and are representative and complete in the sense that major causes of past accidents will be included within the data and that the generic data will represent at least 'average' standards with respect to safety management and safety culture. However, Section 3.3.1 explained that this use of generic data fails to reflect differences in safety standards between different plants, i.e. differences in risk between sites because of differences in systems of management and cultures, and in this sense the assessment is not complete. Chapter 3 provides some important information in relation to these issues, for example:

- Sections 3.3 and 3.4 describe quantified links between risk estimates and standards of safety management, safety culture and organisational structures.
- Figure 3.9 illustrates the idea of a scale of risk performance between 'ideal' and 'real' sites in relation to the theoretical perspectives which describe organisational structure, safety management and safety culture which in turn are derived from Sections 2.1.3–2.1.7.

The purpose of this section is to develop these issues further and to show how measurement of risk differences between sites improves the completeness of the risk assessment and hence its objectivity and credibility. The main points are illustrated in Figure 4.3, which further develops Figures 2.6 and 3.9.

The features of Figure 4.3 are:

- 'Ideal' organisations with low risks.
- 'Real world' organisations with higher risks.
- Organisational theories, or descriptions, of 'idealness'.
- Organisational descriptions of 'real world' organisations.
- Strong overlap between these respective organisational descriptions.
- A scale of risk or accident performance. Real world organisations occupy a position on the scale.
- The idea that the greater the discrepancy between the 'ideal' and the 'real' the greater will be the difference in risk between the sites.
- Generic failure rate data representing an average, or mid-range position, on the risk scale.

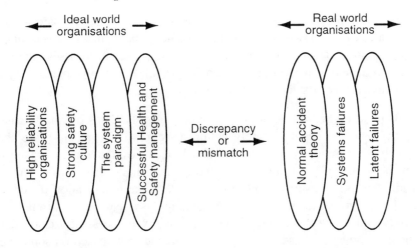

Theories of 'idealness'

- High reliability theory
- Strong safety culture
- The system paradigm
- Successful Health and Safety management

Theories of the 'real' world

- Normal accidents
- System failures
- Latent failures

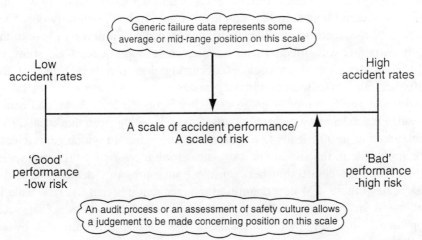

Figure 4.3 *A scale of accident performance (a scale of risk) related to organisational characteristics. Ideal organisations are described by theories of idealness. These organisations have low accident rates (low risk). Real world organisations show a significant difference from the ideal. They have higher accident rates and higher risks. The greater the difference between the ideal and the real world the greater the risk difference, provided technological considerations are equivalent. An audit process allows a judgement to be made about the position of an organisation on the risk scale. This is because the deviations from the ideal are what drives the faults and problems within real organisations. These deviations are measured by the audit process*

- Audit processes (safety management assessments or safety culture assessments) which allow a judgement to be made concerning the position of an organisation on the risk scale, i.e. to assess the extent to which generic data are applicable or not applicable to any risk assessment for the site in question.

Thus in Figure 4.3 audit processes to assess standards of safety management, or assessments of safety culture (Sections 3.3. and 3.4), are shown which are essentially aimed at establishing a position for an organisation on this risk scale. An alternative way of describing this is to say that the audit process measures or judges how 'ideal' the organisation under study is. For example, is the safety management system weak or strong? Is the safety culture weak or strong? The philosophy is the same irrespective of the assessment tool used: the tool is considered to be based on some theory of best practice or 'idealness' and to 'measure' how close the organisation under scrutiny is to this perception of 'idealness'. The audit process measures deviation from 'idealness' and therefore by implication a particular point on the risk scale.

As noted in the discussion to Chapter 3, however, the precise nature of the relationship between the assessment tools (Sections 3.3 and 3.4) and the theoretical descriptions of systems and cultures (Sections 2.1.3–2.1.7) is not always transparent. Chapter 1, case study 10, described failures of management in terms of different levels of management responsibility and failures of communication between them. Case study 11 similarly describes failures of management at all levels in the company: "From top to bottom the body corporate was infected with the disease of sloppiness." Case study 12 illustrates the idea of an accident occurring due to a breakdown of the safety culture. Assessments of safety management system strength or safety culture strength, attempt to judge explicitly issues such as these, and hence to judge the extent to which an organisation is deviating from the 'ideal'. This measurement in turn leads to a measure of the extent to which generic data are applicable to the site in question, measuring a position on the risk scale.

To make the connection between these measures of 'idealness' and the risk scale (Figure 4.3) in a quantified manner requires that the risk scale is calibrated in relation to the various theories of 'idealness'. One such approach,[15] has calibrated the scale of risk shown in Figure 4.3 by an analysis of the range of accident performance actually achieved in practice by a number of different companies and then attempts to measure 'idealness' by reference to sociotechnical descriptions of accident causation, (Section 2.1.5), successful health and safety management (Section 2.1.4) and strong safety culture (Section 2.1.3), i.e. by reference to the range of organisational theories which describe 'idealness'. The analysis of site differences in accident performance data indicated a range of ± 1 order of magnitude (a factor of 100) for the range in performance of plant failure rate data[15] between the 'good' performance end of the scale and the 'bad' performance end of the scale.

This shows that the use of generic failure rate data may fail to show a difference of a factor of 100 between risks for the very best and the very worst companies, depending on the standards of safety management, safety culture and organisational structure. Note that in Figure 3.6 this range of performance between average and very good is a factor of 17.5, which implies a range of about 300 between the very best and the very worst on the scale of risk. This implies that the range of 100 between the very best and the very worst companies may be an underestimate and that failure to include this potential range in the risk assessment introduces a serious omission. The assessment is not complete, and may be in error by a factor of 100 or more. Under these circumstances the risk assessment may be criticised as being subjective and its credibility is potentially undermined.

This section has emphasised the measurement of standards of safety culture and safety management systems in relation to risk and risk assessment because these issues have been shown to be important areas of consideration especially in relation to issues of completeness, subjectivity and credibility and yet their inclusion in risk assessment is in no sense routine.

The case studies from Chapter 1 (case studies 10, 11, 12) which refer to safety management and safety culture clearly show that over and over again these issues are raised in relation to accident causation, and yet risk assessments commonly omit their detailed consideration. The Royal Society Study Group report[16] described this omission as a crucial 'blind spot' in current risk assessment practice. They argued that there were sharp limitations in the then current (1992) state of knowledge about how risk is handled in organisations and called for a more robust knowledge base. They described the risk research map as a bit like the population of Australia, with almost everything clustered round the edges and hardly anything in the central conceptual areas.

Because completeness is a crucially important issue in maintaining the credibility of risk assessment processes, the relationships between safety management, safety culture and risk need to be more firmly established. As Section 4.1.5 shows, risk estimates may be hugely in error, and lack of completeness in the estimation process (Figure 4.2) may be such that whole causal areas are not addressed, so the criticism that the assessment is subjective becomes hard to refute.

4.3 SUMMARY AND DISCUSSION

This chapter has illustrated how estimated risk values are used as an aid to decision making. Against the background of the risk assessment process (risk estimation and risk evaluation) it is clear that the human dimension enters into all the stages, steps and perspectives of risk-based decision making:

- People estimate risks and build their views, biases and judgements into the process of estimation. The risk estimation process may be incomplete and therefore subjective.
- People evaluate risks. Their needs, expectations and values are expressed in the evaluation of the significance of the risk. There are no universal values for acceptable risk.
- People are implicated in the causes of accidents, i.e. they are related to the realisation of hazards both as direct causes of accidents and as underlying causes of accidents.
- People have to 'live with', tolerate or accept risks. They may be involved with decision-making processes or feel that decisions are made on their behalf. They may not feel the decisions are fair.
- Risk assessment as an aid to decision making is often only one aspect of the decision-making process. As such, other issues may in practice dominate. Such issues include, for example, the creation of jobs relating to a new industrial activity, i.e. there is a strong political dimension to risk assessment.

Examples of risk-based decisions are given which illustrate that care is needed to address these human dimensions to risk assessment and that a failure to do this can undermine the credibility of the whole risk assessment process. Nevertheless, risk remains an important and useful concept because it gives the promise of rational decision making, while other concepts such as hazard ranges do not lead to simple decision making.

Particular emphasis is placed on issues of completeness in risk assessment. The failure to address organisational factors in risk assessment is described as a crucial blind spot[16] and can lead to a very significant omission from the risk assessment process. This in turn means that the risk assessment can be described as subjective, incomplete and misleading. For example, the omission of considerations regarding standards of safety management and safety culture are shown to account for difference factors of a 100 or more in estimated risk values and it is hard to refute the criticism that a risk assessment is subjective when whole areas of accident causation are not addressed in the assessment. Within this context the chapter suggests that risk assessments for land-use planning in the UK may be further strengthened if the process addresses issues such as the standards of safety culture and safety management. Approaches to including such issues in risk assessment have been described.[15,17] These approaches allow different standards of safety management and safety culture to be included in risk assessment procedures under appropriate circumstances, although the precise nature of these assessment tools are not yet clearly linked to all the theoretical descriptions which are relevant to them.

Finally, this chapter also shows that in using estimated risk values in decision making care is needed about poorly defined risk concepts, such as those embodied in risk-ranking approaches which are useful only within the context of ranking a range of hazards within a site in some order of 'risk'. This points to the need for carefully defined and used terminology in the risk assessment field. For example, it is important to distinguish between risk ranking on a single site and different risk performance between sites.

FURTHER READING

1. V. T. Covello, *Uncertainty in Risk Assessment, Risk Management and Decision Making*, Plenum Press, New York, 1987.

REFERENCES

1. J. Kadvany *Health, Safety and Environment* 1995, **6**, 333.
2. M. Everley, Health and Safety at Work, October/November 1994.
3. Health and Safety Executive, *The Costs of Accidents at Work*, HSE Books (HS(G)96), HMSO, London, 1997.
4. Health and Safety Executive, *Good Health is Good Business*, HSE Books, HMSO, London, 1996.
5. Health and Safety Executive, *Risk Criteria for Land-use Planning in the Vicinity of Major Industrial Hazards*, HMSO, 1989.
6. Health and Safety Executive, *Quantified Risk Assessment – its Input into Decision Making*, HMSO, London, 1989.
7. N. W. Hurst, C. Nussey and R. P. Pape, Development and Application of a Risk Assessement Tool (RISKAT) in the Health and Safety Executive. Chem. Eng. Res. Des. 67.362. 1989.
8. G.A. Clay, R. Fitzpatrick, N.W. Hurst, D. Carter and P. Crossthwaite, J. Haz. Mat., 1988, **20**, 357.
9. A. I. Glendon, in *Risks and Decisions*, ed. W.T. Singleton and J. Hoven, J. Wiley, London, 1987 p. 101.
10. J. Hoven, in Risks and Decisions, ed. W. T. Singleton and J. Hoven, J. Wiley, London, 1987, p. 162.
11. K. S. Shrader-Frechette, *Risk and Rationality*, University of California Press, Berkeley, 1991, ch. 4. and notes and references therein, especially notes 2–6, pp. 242 and 243.
12. G. W. Bassett, Jr., H.C. Jenking-Smith and C. Silva, *Risk Analysis*, 1996, **16**, 309.
13. V. T. Covello and M. W. Mark Hoffer, *Risk Assessment Methods*, Plenum Press, New York, 1993, ch. 6.
14. Y. Y. Haimes, *Risk Analysis*, 1991, **11**, 169.
15. N. W. Hurst, J. Loss Prev. Process Ind., 1996, **9**, 161.
16. Royal Society, *Risk: Analysis, Perception and Management*, Report of a Royal Society Study Group, Royal Society, London, 1992, p. 181.
17. N. W. Hurst, J. Loss Prev. Process Ind., 1997, **10**, 63.

CHAPTER 5

Discussion and Conclusions – Where Does All This Leave Risk Assessment?

5.1 CONCLUSIONS FROM THE PREVIOUS CHAPTERS

This book has reached a series of conclusions in considering risk assessment. These are now summarised as a series of bullet points. Each of the conculsions is numbered [C1–C30] and the subsequent discussion draws explicitly on the conclusions by back reference to each conclusion number. The purpose of this is to provide as much support as possible for each of the final conclusions reached and to provide a basis for an answer to the question "Where does all this leave risk assessment?" The chapter concludes with an agenda for risk assessment over the next few years which attempts to describe important areas for development of risk assessment firmly based on the conclusions of the book.

5.1.1 Chapter 1

- Illustrates different perspectives (safety engineering failures, human errors and failures of systems and cultures) as the causes of accidents and incidents (the three perspectives). [C1]
- Distinguishes between direct causes of accidents and underlying causes. [C2]
- Distinguishes issues such as the sequences of an accident and how it happened from those relating to the causes. [C3]
- Illustrates the idea of completeness in understanding accident causation which is taken to mean a full consideration of the three different perspectives. If an accident analysis looks at only one area it is likely to be deficient and incomplete. [C4]
- Shows that the issues of completeness are reflected in comprehensive incident investigation reports which hope to prevent similar incidents in the future by learning lessons from the past. [C5]

5.1.2 Chapter 2

- Briefly reviews the theoretical approaches which have been developed to provide a better understanding of accidents and incident causation from the three perspectives. [C6]
- Illustrates that most accidents are characterised by multiple causes. Hardware failures and human error are often the last 'symptom' of underlying problems such as poor planning and inadequate resources. [C7]
- Suggests a simple synthesis (overlap) between descriptions of 'ideal' organisations and 'real' organisations. There is a strong degree of overlap and similarity between some of these approaches. The ideas of a strong safety management system, a strong safety culture, a system paradigm and a high reliability organisation show commonalities and may be strongly interrelated. All these relate to the characteristics of the organisations. Similarly there are overlaps and commonalities between descriptions of 'real world' organisations: system failures, latent failures and normal accidents. [C8]
- Describes the complexity of risk as a concept under the broad heading of a subjective/objective debate. This debate refers to the position of positivists (objectivists) and cultural relativists (subjectivists). These positions represent extremes which describe risk as an objective reality or as a culturally defined concept. [C9]
- Suggests risk is neither subjective nor objective absolutely. [C10]
- Suggests that despite difficulties in defining risk, activities using similar technology poses different risks to the people involved with them. This is an objective issue. Thus, despite the uncertainty about the definition of risk, its cultural dimension and the difficulty in estimating risk, it is true that two individuals doing similar jobs in different places may run different risks. Because of this, the estimation and management of risk are important issues. [C11]
- Expresses caution about the interpretation of risk perception and risk communication studies when used in an attempt to establish right and wrong with respect to decisions about acceptable risk. [C12]

5.1.3 Chapter 3

- Introduces the links between the three perspectives and quantified risk estimates: [C13]
  - risk and engineering
  - risk and people
  - risk and systems and cultures
- Suggests five distinct reasons for caution in acceptable risk problems: [C14]
  - issues of completeness
  - the subjective element in risk

- the absence of universal values for acceptable risks
- risk is often part of a wider debate
- who benefits and who bears the risk?
- Illustrates a range of 'relative objectivity' for risk assessments. Some risk estimates are more objective than others. This relates to the extent to which evidence enters into the estimation process. [C15]
- Links the descriptions of 'ideal' and 'real world' organisations with the idea that activities using similar technology pose different risks to the people involved with them. [C16]
- Suggests it should be valid to compare risks between different sites in terms of safety management, organisational structure and safety culture differences provided that technical considerations are broadly equivalent. [C17]

5.1.4 Chapter 4

- Concludes that the human dimension enters into all the stages of a risk assessment process. [C18]
- Suggests that the concept of risk nevertheless holds the promise of rational decision making. [C19]
- Illustrates that the consideration of hazard ranges for chemical leaks does not lead to a simple decision-making process. [C20]
- Distinguishes between quantified risk assessment and risk-ranking procedures. [C21]
- Distinguishes between risk ranking on a single site and different risk performance between sites. [C22]
- Illustrates a decision-making process based on acceptable risk for land-use around major hazard sites in the UK. Suggests that the five areas of caution [C14] (Section 3.1.1) are adequately considered in the process. [C23]
- Contrasts the decision making concerning land use around major hazards in the UK with attempts to find suitable sites for the disposal of radioactive waste. Suggests that in some circumstances acceptable risk decisions cannot be made. [C24]
- Shows how measuring risk performance between sites can further strengthen such risk-based decisions for land use planning by improving the completeness of the process and addressing a key area of subjectivity. [C25]
- Shows that the credibility of decision making based on estimated risk levels can be undermined if such issues as completeness and subjectivity are not clearly addressed. [C26]
- Supports the view that the limited attention so far paid to the influence of organisational and management factors on risk is a critical blind spot in current risk assessment practice. [C27]

- Mentions and refers to the development of methods to address this blind spot (Chapter 3). [C28]
- Suggests that it is hard to refute the criticism that a risk assessment is subjective when whole causal areas are not addressed. [C29]
- Suggests the need for carefully defined and carefully used terminology in the risk assessment field. [C30]

Having listed the main conclusions from the previous chapters, the purpose of the remaining sections of the book is to ask the question "Where does all this leave risk assessment?" and to use the conclusions to support the answers and suggest an agenda for the future. The reader is invited to refer back to the relevant conclusion.

5.2 RISK ASSESSMENT – THE HUMAN DIMENSION

This book has established that risk assessment has a strong human dimension [C1, C13, C14, C18]. This is illustrated in Figure 5.1 which shows the involvement of people throughout the risk assessment process.

Chapter 4 summarised these involvements:

- people estimate risks
- people evaluate risks
- people are implicated in the causes of accidents
- people have to live with, accept or tolerate risks
- other important human considerations (economic, political) enter into risk assessments

RISK ASSESSMENT	
Estimate the risk - How big is it?	Evaluate the risk - How significant is it? Is the risk acceptable?
The Human Dimension Pervades Risk Assessment	

Figure 5.1 *Risk Assessment – the human dimension. The figure illustrates how people enter into all stages of risk assessment. This is not to imply that risk assessment is a linear process involving an estimation stage (objective) and an evaluation stage (subjective), but to show that people are involved with all the various steps of risk assessment*

It seems unnecessary to labour the point, but clearly the concept of risk assessment as a pure, objective, scientific activity needs to be abandoned. [C9, C15]. This applies equally to the estimation of risk and to the evaluation of risk. This point is made by many authors. For example, Dale Jamieson,[1] in discussing uncertainty in relation to risk says: "... uncertainty is constructed both by science and by society in order to serve certain purposes. Recognising the social role of scientific uncertainty will help us to see how many of our problems about risk are deeply cultural and cannot be overcome simply by the application of more and better science." So risk assessment is not a pure science and one aspect of its human dimensions is expressed through the debate concerning the objective or subjective nature of risk assessments. This debate is further developed below.

5.3 THE SUBJECTIVE/OBJECTIVE DEBATE

In her book *Risk and Rationality*[2] K. S. Shrader-Frechette characterises the subjective/objective debate in risk assessment in terms of two extreme positions represented by the 'cultural relativists' and the 'naive positivists'. These extremes represent the two positions which may be summarised as defining risk as a subjective and socially constructed idea (cultural relativism), or as an objective and measurable reality (naive positivism). Both of these extreme positions are criticised. To quote from the book: "... both of these frameworks, naive positivism and cultural relativism, err in being reductionistic. The cultural relativists attempt to reduce risk to a sociological construct, underestimating or dismissing its scientific components. The naive positivists attempt to reduce risk to a purely scientific reality, underestimating or dismissing its ethical components." Furthermore, both of these positions are criticised for being antipopulist because both are critical of lay evaluations of technological and environmental risks. The cultural relativists accuse the public of being superstitious while the naive positivists accuse the public of inconsistency.

> Cultural relativists begin from an astute (although not original) insight. They recognize that risk estimation is not wholly objective and criticise risk assessors for their repeated error in assuming that lay estimates of risk are mere 'perceptions' whereas expert analyses are 'objective'. Clearly the cultural relativists are to be commended for discussing this error, since both experts and the public have risk 'perceptions' and no one has privileged access to the truth about risk. Cultural relativists err, however, in assuming that, because everybody can be wrong at some time, everybody is wrong all the time.

Shrader-Frechette argues that the cultural relativists make too strict demands (on risk estimation) by presupposing that 'objective' risk estimates must be wholly value free and therefore infallible and universal. She suggests that risk assessments are 'objective' or 'scientific' in a number of important ways:

- they can be the subject of rational dispute and criticism.
- they are partially dependent on probabilities that can be affected by empirical events.

Where Does All This Leave Risk Assessment? 85

- they can be criticised in terms of how well they serve the scientific goal of explanation and prediction.

Because of this she argues that although risk assessment is subject to various shortcomings and biases it should continue to be used. It needs to be improved and used in policy making. Although risk assessment is in practice flawed, it is in principle necessary for rational, objective, democratic assessments of risk.

Thus she proposes a path between these two extreme positions: an account of 'rational risk evaluation' which has its main policy goal to enfranchise the very people who are most likely to be victimised by environmental hazards. She calls this midway path 'scientific proceduralism'. Crucial of this midway path are the extent to which risk assessments are objective, as listed above, and the need for informed consent and citizen participation in negotiating solutions for problems of risk, especially where one group of citizens might receive the benefits associated with a hazard while another set of persons might bear the costs. This is suggested as a practical way to avoid the NIMBY (not in my backyard) syndrome. It is not suggested that negotiation is easy, but it is necessary. These ideas are further developed below, where the implications of the conclusions reached in this book are also discussed for both risk estimation (Section 5.4) and risk evaluation (Section 5.5).

However, it is important to realise that the subjective/objective debate concerning risk assessment is taking place within a far wider context which involves other subject areas, for example medicine, economics and the law, and all of the areas are influenced by thinking concerning objectivity and social relativism. There is a wider social debate about rules and objectivity, versus feelings and values. This wider debate and context are briefly discussed before implications for risk assessment are developed.

5.3.1 The Wider Context

Perhaps the most famous *cause célèbre* which illustrates the wider context within which the risk debate is now taking place is the 'Sokal hoax'[3] [C9]. In the autumn of 1994, a New York University theoretical physicist, Alan Sokal, submitted an essay to *Social Text*, a leading journal in the field of cultural studies. Entitled 'Transgressing the Boundaries: Towards a Transformative Hermeneutics of Quantum Gravity', the essay purported to be a scholarly article about the 'post-modern' philosophical and political implications of twentieth-century theoretical physics. However, as the author himself later revealed in the journal *Lingua Franca*, his essay was merely a hotchpotch of deliberate blunders, howlers and non-sequiturs, stitched together so as to sound plausible.

Paul Boghossian[3] believes that the hoax shows three main points:

- that dubiously coherent relativistic views about truth and evidence really have gained wide acceptance;
- that this has had pernicious consequences for standards of scholarship and intellectual responsibility;
- that neither of the preceding two claims need reflect a particular political point of view.

He argues that scientific evidence and facts must remain a key part of scholarship, rationality and understanding. He is supported by many others. For example, in an obituary to Thomas S. Kuhn, the author of the famous *Structure of scientific revolutions*, David Hull[4] attacks the use of Kuhn's work by social relativists. Kuhn had argued that scientific 'paradigm shifts' cannot be explained entirely in terms of reason, argument and evidence. This has been interpreted by relativists to mean that paradigm shifts are unrelated to reason, argument and evidence. This seems to be a recurrent theme. It is argued that because facts are uncertain, or evidence incomplete, facts and evidence have no part to play in decision making or rational behaviour [C10] [C15].

Another account of this 'wider context' is provided by a recent article in *Scientific American*.[5] There it is argued that science is 'getting at' physical phenomena but this is not to reject the idea that culture can influence science [C15]. It is suggested that science should aim for 'strong objectivity' – a means of evaluating not only the usual scientific evidence but also the social values and interests that lead scientists towards certain questions and answers.

Clearly the objective/subjective debate within risk assessment is closely related to this wider context. The debate has the potential to become very polarised and adversarial. But as Shrader-Frechette describes, a middle ground is both desirable and achievable in the risk assessment area. However, this middle ground needs to recognise the strengths of both sides of the debate and to devise solutions accordingly.

5.3.2 Comparisons With Other Areas – Medicine, Economics, Law

Because of the broad context within which the risk assessment debate is taking place it is not surprising to find that other areas of activity in which evidence has a part to play are also concerned with similar issues. I wish to mention some of these briefly.

In the area of medicine, risk is a widely used concept. The terminology of risk is used to describe surgical procedures and drugs. For example, an operation may have a 1:500 mortality rate. Or a particular drug may cause adverse side effects in 10% of patients. The acceptability of these risks depends on individual circumstances. There are no universal values for acceptability. For an individual facing an illness whose outcome is certain death, a procedure with only a 50% chance of success may be judged acceptable. It would be a personal choice. Of course, the statistics will be uncertain because they relate to averages not individuals and, furthermore, medical knowledge is not absolute but represents the best available at that particular time. Most people, however, faced with a medical problem will accept that the current state of medical knowledge and evidence is the best basis on which to make a decision on their need for treatment. This is difficult decision making in the face of great uncertainty.

In economics, modelling of economic outcomes is a normal part of the 'art'. The economic models make use of experience, and supposed relationships between, say, inflation and interest rates, wages and unemployment and so forth. One purpose of economic modelling is to help judge levels of interest rates, taxation and spending. Obviously, the whole process is political and highly uncertain but nevertheless it is informed by evidence and experience. Some doubt the value of such economic modelling because of the uncertainties, but it remains an integral part of economic prediction.

In law, a magistrate or jury will judge the guilt of a defendant based on the balance of evidence. The magistrate cannot be sure that the defendant is guilty but, must form a judgement taking together the evidence both for guilt and innocence. The evidence may well include scientific evidence, e.g. concerning DNA or other forensic information. This is clearly a process in which evidence is used to inform a judgement, but the evidence can never be 100% uncontroversial. For example, the DNA results may show a 1:40,000 chance that the defendant is innocent. Is this sufficient evidence not to convict? Perhaps the answer depends on – guilty of what? The balance of evidence is not so crucial if conviction implies a £50 fine or a life sentence. This again represents decision making in the face of uncertainty.

So risk assessment is part of a wider debate – is it a science and if not, how objective or subjective is it? In turn, many other areas of human activity, such as medicine, economics and the law, are all similar. Facts and evidence are important but uncertainties are high. Nevertheless, decisions must be made. Against this background, the implications of the book's conclusions will be developed for risk assessment.

5.4 IMPLICATIONS FOR RISK ESTIMATION

A clear implication for the practice of risk estimation which emerges from this book is that of the pursuit of completeness and balance. [C4, C5, C26, C27, C28]. Figure 5.2 again illustrates the major causal areas which need to be addressed in risk assessments.

In the developing field of risk assessment, these three areas have received different amounts of attention and represent different levels of sophistication [C27]. Reliability engineering applied to hardware failures,

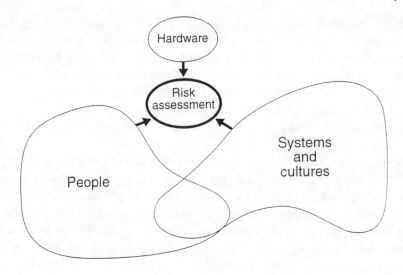

Figure 5.2 *Risk Assessment – balance and completeness. Three major causal areas need to be considered with similar weight and emphasis*

different design options, modelling of fires and explosions, the dispersion of dense toxic gases and so forth have all attracted considerable attention and effort and are well developed [C6, C13]. There is a strong theoretical basis for the work involving engineering design, reliability engineering, fluid mechanics and so on. Similarly the effects of people, as individuals, on the reliability of engineering systems has been well described via ergonomics, task analysis, human reliability assessments and logic trees. Again there is a strong theoretical basis [C6, C13] for the work (although psychological theory suggests limits to the extent to which the effects of people on systems can be modelled).

The area of work described under the heading of systems and cultures would seem to be far less well developed [C13]. There are two aspects to this. First, there are competing and disparate accounts of organisational functioning [C8]. Scott Sagan's book, *The Limits of Safety* (see Section 2.1.7) sets out this controversy very clearly. Furthermore, various other organisational descriptions seem to overlap strongly. This is shown again in Figure 5.3.

There seems to be a need for a unifying approach to these various theoretical descriptions of organisations for use within the context of risk assessment and risk management. Secondly, the concept of linking risk to organisational functioning is not well established. In this book, approaches to this problem are described as 'experimental' [C13, C28], and although the concepts of safety culture and safety management are considered important the link with risk is not well established [C13].

From the point of view of completeness and balance in risk assessment

Where Does All This Leave Risk Assessment?

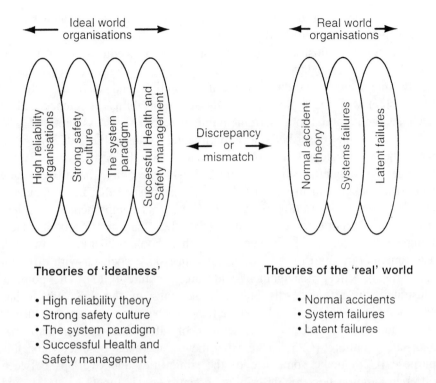

Figure 5.3 *There are disparate and competing accounts of organisational functioning. Often these overlap and this suggests the need to develop unified theories which can embrace these different approaches*

the disparity between the different stages of development and application of the three causal areas represents a considerable problem. Risk assessments which exclude a consideration of safety management systems and safety culture clearly lack balance and are incomplete [C25, C26, C27]. This is the crucial 'blind spot', and such incompleteness and lack of balance has the potential to undermine the credibility of the entire risk assessment process [C29].

This position is supported by other workers. For example, William Freudenburg[6] writes: "The public trust is valuable, but it is also fragile, and it is highly susceptible to corrosive effects of scientific behaviours that fall short of the highest standards of responsibility"; and David Blockley[7] writes: "Two world-views of hazard, risk and safety with reference to the engineering industry are discussed: the one technical and the other human and organisational. It will be argued that there is a need to unify these two views ... The simple message that emerges is that safety depends on good management." Similarly Clive Smallman[8] writes: "Our record in identifying potential human and organisational failures, exacerbated by political factors, is less good. Evidence of this exists in a string of recent major tragedies and disasters where alleged human and organisational failures led to loss of life and property."

These quotes illustrate that risk estimation procedures are often incomplete and lack balance because of a failure to address the key areas of safety management systems and safety culture and also that this is inconsistent with the highest standards of responsibility in science and as such has the potential to undermine public trust in risk assessment.

This book has described the experimental systems [C13, C28] which have been developed to make explicit the links between safety management standards, safety culture and risk. These links have been made against a background in which it has been established that most accidents are characterised by multiple causes [C7] and that comprehensive incident investigation reports try to establish all of these causes [C5] both underlying and direct [C2]. Nevertheless, this book illustrates how risk assessments often fail to reflect the different risks associated with different standards of safety management and safety culture [C17]. The point is illustrated by reference to land-use planning in the UK which has considerable strengths [C23]. Nevertheless, the process can be further strengthened by an explicit consideration of safety management standards and safety culture [C25]. The experimental methods described for this purpose [C28] make some use of the similarities and overlaps between different approaches to describing organisational functions [C8] and crucially the idea that activities using similar technology pose different risks to the people involved with them [C16]. It is suggested that it is valid to compare risks between sites in terms of safety management, organisational structure and safety culture differences provided that technical considerations are broadly equivalent [C17].

Another general implication for risk estimation is concerned with the type of language which is used to describe the risk assessment. Because of the subjective nature of risk assessment [C9] care is needed to use appropriate descriptions of the process and its outputs. But risk is not a uniform concept [C15]; some estimates of risk are more objective than others. Because of this, the language which is used in risk assessment needs to reflect the relative objectivity of the risk estimate [C15]. This is illustrated in Figure 5.4 which builds upon the subjective/objective representation of risk assessment shown in Figure 3.8.

Figure 5.4 suggests that if an estimate of risk is strongly based on evidence, experience, statistics and so on then it is appropriate to describe the estimated risk in terms of that evidence, i.e. to describe the evidence, statistics, logic and modelling which have been used to arrive at the estimate. On the other hand, if the risk estimate contains a large subjective element, e.g. there is lack of good statistics and lack of experience, then the risk assessment needs to be framed differently and to use the language of judgement, views and possibilities. The issue of appropriate language

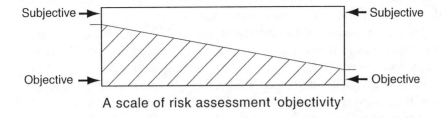

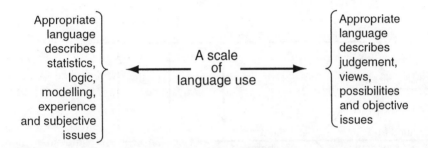

Figure 5.4 *A scale of risk assessment objectivity. Some risk estimates are more objective than others. Similarly it is appropriate to use language which is consistent with the 'objectivity'*

clearly relates to issues of trust (see above) and risk communication. These are discussed in the next section.

Finally, in considering implications for risk estimation the issue of clear statements and use of definitions needs to be considered. This does not necessarily relate to a need for all workers to use the same definitions, simply for workers to define their terms, understand other definitions and to be clear about any differences [C21, C22, C30]. For example, we must be clear about definitions of risk [C21], and be able to distinguish between risk ranking on a single site and risk comparisons between sites [C22].

5.5 IMPLICATIONS FOR RISK EVALUATION

5.5.1 New Frameworks

Shrader-Frechette,[2] in considering the implications of her work for risk evaluation (see Section 5.3), calls for 'rational risk evaluation' to make the best use of the objectivity of risk assessment and to provide a mechanism for public involvement and consent in negotiating solutions to problems of risk. This theme has been extensively considered and developed by others.[9-12] The National Research Council in the USA has proposed a 'new framework' for assessing environmental and public-health risks, allowing the public and scientists to become involved in discussing risks before they are given a formal assessment.[10] The proposal is designed to have far reaching implications for the way in which the US Government uses science to reach decisions about managing such risks.

Their proposal is illustrated in Figure 5.5 which is reproduced in a simplified form from Reference 10. The approach emphasises the need "to get the right science, and to get it right and to get the right participation and to get that right", i.e. to use appropriate scientific methods in a careful way, and to involve relevant persons in the formulation of the risk problem and discussions about acceptability.

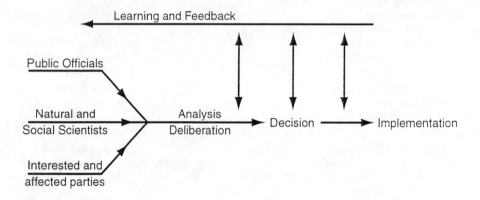

Figure 5.5 *A new risk assessment framework which involves the public from the start. It represents a new approach for the USA suggested by the National Research Council, for risk-based decision making. The new framework would allow the public, scientists and public officials to become involved in discussing risk before they are given formal assessments*
(Modified from Ref. 10).

A key feature of the discussions is public involvement, via interested and affected parties, particularly to develop and maintain trust in the risk assessment process. For example, in Reference 11 the editors say that recognising the problem of trust as a key factor in understanding risk perception and risk conflicts has vast implications for how we approach risk management in the future. Creating and maintaining trust may require a degree of involvement with the public that goes far beyond public relations and two-way communication. It suggests levels of power sharing and public participation in decision making that have rarely been attempted.

The acceptability of the involvement of a far wider public is dependent upon officials and risk assessors abandoning the notion of risk estimation as a pure science and of risk assessment as a linear process involving an objective estimation stage followed by a subjective evaluation stage. For example, Funtowicz and Ravetz[13] say that with the recognition of its craft and informal aspects as essential elements, risk assessment and management abandons its mantle of being an applied science or an engineering task, rejoining the family of human endeavours. It is one of those emerging sciences where, typically, facts are uncertain, values in dispute, stakes high and decisions urgent. Funtowicz and Ravetz refer to this family of human endeavours as 'post normal' sciences. The examples of the law, medicine and economics would seem to fit in well with this description, as does risk assessment. It is only when risk assessment is considered as a post normal science that these new frameworks, or approaches, make sense as an approach to risk-based decision making.

The wider involvement of a public of affected and interested parties is very controversial.[9] It raises questions concerning whom the new participants would be, how the interactions would be managed and how conflicts would be resolved. The question of these conflicts is addressed directly by Christopher Hood[9] by considering approaches to risk management based on institutionalising rival values and keeping them in opposed tension, so that value clashes become explicit.

5.5.2 Language and Communication

In Section 5.4 above, the issue of language use in risk assessment is raised in particular in relation to a scale of relative objectivity in risk estimation (Figure 5.4). This issue relates strongly to issues of communication in new frameworks for risk assessments, but there is a potential problem in the nature of communications within an extended new framework.

This book has expressed concern about the use of concepts related to risk perception and risk communication in an attempt to establish right and wrong with respect to decisions about acceptable risk [C12]. Others, e.g. Mary Douglas,[14] argue that the developing risk amplification and risk communication approaches also fit with an approach to risk assessment which sees it in terms of a linear objective (estimation) and subjective (evaluation) process. Concepts of social amplification of risk, of risk communication and extended public involvement are then seen as an attempt to allay irrational public fears and defuse those issues most likely to give rise to social concerns. However, as mentioned above the new frameworks are predicated on abandoning such views of the scientific objectivity of risk estimation. Without this there is a danger that new frameworks will be pursued and developed by different people, or groups, who have an entirely different view of what risk assessment is and how the new framework will eventually operate in practice, and what it aims to achieve.

It would seem that the best use of the insights provided by the risk perception and risk communication literature would be within these new frameworks based on the notion of risk assessment as a post normal science.

5.6 FINAL DISCUSSION

This book has concluded that the concept of risk holds the promise of rational decision making [C19] because, among other things, despite all the difficulties associated with risk, similar technologies pose widely different risks to the people affected by the technology [C11]. This conclusion supports the need for continuing to use risk assessment procedures but with better understanding of the strengths and weaknesses of the risk assessment. How might this better understanding evolve?

I believe that the notion of risk assessment as a post normal science needs to be further accepted. This does not mean that I accept the extremes of the subjective argument that risk is an entirely constructed and subjective idea formed from the minds of people. Indeed, I have been at pains to describe the extent to which risk is objective. But others argue that risk assessment is a pure and an objective science, and this description seems to be increasingly untenable. The description of risk assessment as a post normal science seems much more appropriate. Similarly, the ideas of new frame- works for risk assessment (risk estimation and risk evaluation) seem encouraging, but I believe that risk assessors (those who estimate risk) need to look more openly at the methods which are currently used to estimate risk. In considering the new frameworks proposed for risk assessment (Figure 5.5) it is hard to be optimistic about their use at the present time when:

- the proposed frameworks are likely to be misinterpreted in their intent by whose who believe that risk assessment is an objective science, and
- risk estimation is so fraught with conceptual and definitional problems, such as lack of completeness, lack of balance and confusion about the meaning of key terminology.

Taking these points one at a time; I think there is still a strong view that risk assessment is essentially objective. This view sees the literature on risk perception and risk communication as an extension to the 'objective risk assessment view' and would easily absorb the new frameworks into this view, seeing the frameworks as a way of allaying irrational fears about 'safe' technology. Secondly, risk estimation is so fraught with factual, conceptual and definitional problems that these new frameworks are sure to expose these weaknesses. I have particularly emphasised weaknesses that relate to issues of completeness and balance in risk assessment. Thus, although in descriptions of accidents and in reports of incident investigations the pursuit of completeness is often considered highly meritorious, when risk estimation is concerned no such completeness is found. Yet this is highly illogical because risk assessment is supposed to answer questions such as "What can go wrong?" and "What are the likely consequences?" What better source of information than incident investigations can there be to understand the root causes of incidents? To ignore root causes of incidents, demonstrated by accident investigations, shows only that risk assessment is a relatively recent activity which has, as yet, not fully developed and matured. How will risk estimation methods stand up to the exposure of the new frameworks? These emphasise getting the right science and getting it right, but do we yet know what the right science is or how to get it right?

And what of this completeness? Reliability engineering is well developed; next in line of completeness is human error as a direct cause of accidents and, finally, least complete is the relationship between risk and what I have called systems and cultures. This order of development reflects two things. Firstly, it reflects which areas have been most easy to develop and, secondly, it reflects the order of development that is most convenient. Ease does not refer to simplicity because reliability engineering is a complex mathe- matical subject, but it has involved making use of established mathematics and logic approaches which were available. Similarly, the methods of human reliability have built on reliability engineering and human psychology. But the subject area of organisational factors and their effect on risk is not well developed and is still confused.

When I say that the order reflects how convenient it has been to develop these areas I mean that reliability engineering refers to engineering parts and components and how to make them safe. I have tried to explain how important this is and how science and engineering play a vital role in health and safety. But it is not the whole story. Human reliability analysis as applied to individuals 'at the sharp end' is convenient because it extends this reliability analysis to include humans, but these humans often have little influence or power in how the analysis is carried out or the results used. It is convenient because it emphasises human error at the level of the chemical plant operator or the train driver. So it shields the organisations these people work for from criticisms of poor resourcing, poor systems, poor definition of tasks and so forth. This brings us back to the insights of the cultural relativists because the people who have been least exposed by risk assessment are those who have most power and influence in the organisations they run.

The whole field of risk assessment grew out of reliability engineering. But at this point in time (1997) major inputs have been also been provided in the risk debate by psychologists, sociologists, political scientists, decision analysts, philosophers and so forth. Now I believe that it is time for the estimators of risk to grasp again the problems facing risk estimation. As Mary Douglas didn't say: "What's wrong with risk estimation? Not much really, just the methods we use, the models we make and the data we have available!"*

5.6.1 An Agenda for Risk Assessment Over The Next Few Years

A driving force for me in developing the ideas for this agenda has been the range of accident performance which different real organisations achieve although they are using similar technology. I know that statistical comparisons of this kind need to be treated with care and can be subject to distortion or misrepresentation but I also know from my first-hand experience that there are real differences in safety culture and safety management between sites using essentially the same technology, and this is reflected in real differences in the number of accidents which occur on the sites.

Provided this difference in accident performance between sites is considered in a sensible way, by also considering how large the sites are, how many people are employed and the tonnage of product produced, then it is possible to express the difference in accident performance in terms of differences in risk (chance of a specified outcome). This difference in risk on the sites affects the lives of real people who work on a day-to-day basis on the sites. Put in stark terms, some organisations routinely injure their employees while others who carry out essentially similar work simply do not. To me this represents real differences in risk, because the chances of being hurt during a working lifetime in the two organisations are different and can be shown to be different.

There are a number of direct implications of this position. Firstly, it is necessary to be clear and precise about the terminology used in the discussions of risk differences because if the discussions make use of definitions which are confusing and which have different meanings to different people then the force of the argument is lost. It is essential that this does not happen because the argument is important: the chances of being hurt in a working life in two organisations are different.

Secondly, I have devoted a lot of space to describing the subjective/objective debate in risk assessment and to placing the debate in a wider context. I believe that a careful consideration of the risk difference between sites is a powerful rebuttal to the challenge that risk is merely a social construct.

Yet considering risk assessment as a pure science is not helpful either because over and over again what I have called the human dimensions to risk assessment are clear. This is why I believe that those of us who estimate risk must address risk assessment as a post normal science not a

---

*This is a travesty of the famous quote "What are Americans afraid of? Nothing much, really, except the food they eat, the water they drink, the air they breathe, the land they live on, and the energy they use."

Where Does All This Leave Risk Assessment?

pure science. I do not have any problems with this. I have tried to show that other areas of human endeavour, the law, medicine and economics, all operate in a manner which is essentially that of a post normal science. That is to say difficult decisions have to be made, evidence and experience are used to help form the judgements, a wide range of inputs and discussions are welcomed and form part of the evidence and at the end of the day a decision is made. Clearly this does not mean the evidence and experience lead unequivocally to a decision but equally it does not mean the evidence is irrelevant to the decision. This has to be the essence of risk assessment.

Thirdly, I have mentioned appropriate use of language in risk assessment and the concept of 'relative objectivity'. I believe that this is important within the context of risk assessment as a post normal science, because it provides the opportunity for risk assessors to describe their work in terms which express the power of the evidence and yet include the elements of judgement. On the other hand, if the assessment is largely subjective it acknowledges first the judgements which have been made but also includes the evidence and statistics which can be used.

The differences in risk performance between sites using essentially similar technology leads to another important set of ideas. It is the nature of organisations and their contexts which is driving these differences. This is why so much effort has been devoted to safety management systems, safety culture, high reliability organisations and so forth. Equally it is why auditing methods to assess standards in these areas are increasingly used and applied as a normal part of safety management in larger organisations. My observation is that these various descriptions of organisational functioning show areas of commonality and that as a consequence of this a clear challenge for the future is to develop 'unified accounts' of organisational function. Linked to this is the nature of the audit and assessment methods used to assess standards. Plainly these need to be more clearly linked to the accounts of organisation function and if a 'unified account' can be developed then audit methods based on this unified account will also need to be developed.

Finally, in considering these ideas with respect to risk assessment my observation is that risk assessments are often incomplete because they are weak in addressing the area of organisational factors in the assessment. But this is a crucial area of consideration and it is therefore important both that ways of including organisational factors in risk assessment continue to be developed and also that the link between risk and organisational functioning are further explored.

My agenda for risk assessment over the next few years covers all of these areas. I believe the agenda represents a considerable challenge but will greatly strengthen the basis and use of risk assessment because it is based on logical arguments and analysis.

Thus, I believe that those of us who estimate risk must now grasp and take forward some key issues which I call an agenda for risk assessment:

- Risk assessment is a 'post normal' science. It is not purely objective and the human dimension pervades risk assessment. The notion of risk assessment as a pure science must be abandoned. Risk assessors must explore the implications of this for their work.
- There is a wide range of inputs that should form part of the risk debate. These should be welcomed for the insights they bring. Risk assessment should be considered as a multi-disciplinary activity and not the preserve of any one discipline. Again, the implications for risk assessment should be explored via multi-disciplinary teams.
- Risk assessment does have a strong objective element to it. The extent of this objectivity will depend on the particular risk assessment (relative objectivity) but objectivity is a key part of risk assessment and estimators of risk should explain and defend the use of science and statistics in their work. These scientific and statistical aspects should be the subject of processes of continuous improvement and development.
- Key areas of completeness and balance need to be addressed in risk assessment:
   – A unified account needs to be developed to describe 'ideal' and 'real world' organisations for use within risk assessments (Figure 5.3).
   – These unified accounts need to be linked directly to audit methods for the assessment of standards of safety management and safety culture.
   – The relationship between risk and organisational functioning needs to be explored and developed. Explicit links are needed, which develop the use of a unified account of organisational function.
- A key problem area is the definition and use of terms in risk assessment which have become confused and inconsistent. This confusion needs to be addressed.
- Language for use in risk assessment, and in new frameworks, needs to reflect the relative objectivity of the risk assessments. This is a vital part of improving trust in the risk assessment process and in enabling new frameworks for risk assessment to be developed and used in a constructive way.

This short list represents what I believe is an important agenda for risk assessment research over the next few years, and represents my attempt to answer the question "where does all this leave risk assessment?" Risk assessment is under attack and risk assessors need to get their house in order so as to preserve what is good about risk assessment and to improve its areas of weakness. Because risk assessment is a post normal science it is vulnerable to attacks of subjectivity and arbitrariness. The response to this must be to continue to develop the strengths of risk assessment but also to accept its limitations.

REFERENCES

1. Dale Jamieson. The Annals of the American Academy of Political and Social Science. Challenges in Risk Assessment and Risk Management.
   Page 35 – Scientific Uncertainty and the Political Process.
   Vol 545 May 1996. Special Editors, H. Kunreuther and P. Slovic.
2. K. S. Shrader-Frechette, *Risk and Rationality*, University of California Press, Berkeley, 1991.
3. P. Boghossian, *Times Literary Supplement*, London, 13 December 1996, p. 14.
4. D. L. Hull, *Nature*, 1996, **382**, 203.
5. The staff of the Scientific American, *Scientific American*, January 1997, p. 80.
6. William Freudenburg. Risky thinking: Irrational Fears about risk and society. The Annals of the American Academy of Political and social science. Challenges in Risk Assessment and Risk Management Page 44. Vol 545 May 1996 Special Editors. H Kunreuther and P Slovic.
7. D. I. Blockley, in *Accident and Design – Contemporary Debates in Risk Management*, ed. C. Hood and D. K. C. Jones, University College London Press, London 1996, p. 31.
8. C. Smallman, in *Risk Homeostasis and Risk Assessment*, ed. I. Glendon and N. Stanton, (Safety Science, Special Issue Vol. 22, Pergamon, Oxford, 1996, p. 245.
9. *Accident and Design – contemporary debates in risk management*, eds. C. Hood and D. K. C. Jones, University College London Press, London, 1996.
10. *Understanding Risk: Informing Decisions in a Democratic Society*, eds. P. C. Stern and H. V. Fineberg, National Academy Press, Washington DC, 1996.
11. The Annals of the American Academy of Political Social Science.
    Challenges in Risk Assessment and Risk Management. A special issue, ed. Howard Kunreuther and Paul Slovic Vol 545, May 1996.
12. C. Macilwain, *Nature*, 1996, **381**, 638.
13. S. O. Funtowicz and J. R. Ravetz, *Risk Analysis*, 1992, **12** 95.
14. Mary Douglas, 'The Politicisation of Risk and the Neutralisation of Politics', Paper presented to the Political Economy Research Lecture Seminar, University of Sheffield, 17 February 1994.

Subject Index

Acceptable risk, Glossary, 31, 32, 70
Accident rates and safety management, 53
Active failure, 26, 27
Agenda for risk assessment, 96–98
Allied Colloids, 2
Ammonia, 4, 5
Attitude scales, 57
Attitude surveys, 57
Attitudes to safety, 20–23
Audit processes, 76

Balance, 87, 89

Clapham Junction railway accident, 7
CMA AVG, 55
Communication, 93
Completeness, 11, 60, 73, 87, 89
Consequence models, 16, 44
Credibility of risk assessment, 74–77

Economics, 70, 87, 93
Engineered safety, 15
Engineering approaches to accidents, 14–17
Engineering approaches to risk assessment, 43–48

Fault tree analysis, 45
Feyzin, 6

Hardware failures, 2–6, 14–17
Hazard, Glossary

Hazard identification, 16, 17, 43, 44
HAZOP, 17, 44
HEART, 49
Herald of Free Enterprise accident, 10
Hidden report, 7
High reliability theory, 27–29, 32
Hillsborough football stadium, 4
Housing development, 70
Human reliability assessment (HRA), Glossary, 48–50
Human performance, 17–20
Human errors, 7–9, 17–20

Ideal world organisations, 35
IFAL, 52
Intrinsic safety, 15
ISRS, 53, 54

Kegworth air crash, 8

Land-use planning, 70
Language, 90, 93
Latent failures, 26, 27, 35
Law, 90, 93
Lost time injury rate, 55

Major hazard sites, 70
MANAGER, 52
Medicine, 86, 93
Modification of risk, 51–53

New frameworks, 91, 92
Normal accidents, 27–29

Subject Index

Nuclear waste storage, 72

Objective/subjective, 29–31, 36, 37, 84

Parkhurst Prison escape, 10
People failures, 7–9
Post normal science, Glossary, 93
Potchefstroom, 6
PRIMA, 52
Probabilistic risk assessment, Glossary
Probabilistic safety assessment, Glossary
Public involvement, 92

Quantified risk and safety culture, 56, 57
Quantified risk and safety management, 51–56
Quantified risk assessment (QRA), Glossary, 43–46

Real world organisations, 35, 39
Relative objectivity, 59, 91
Risk, Glossary
 as a useful concept for decision making, 65
Risk assessment, Glossary
Risk assessment agenda, 96–98
Risk-based decisions, Glossary, Chapter 4
Risk communication, 93
Risk/cost trade-offs, 70
Risk estimation, Glossary

Risk evaluation, Glossary
Risk management, Glossary
Risk perception, Glossary, 31, 32, 39, 84
Risk ranking, 68, 69

Safety, Glossary
Safety culture, Glossary, 9–11, 20–23, 32
Safety culture and risk, 56, 57
Safety management and risk, 51–56
Safety management (systems), Glossary, 9–11, 23–25
SLIM, 49
Social amplification of risk, Glossary, 93
Socio-political amplification of risk, Glossary, 36
Sociotechnical systems, 25, 26
Sokal hoax, 85
Subjective/objective, 29–31, 36, 37, 84
System failures, 9–11, 33, 34
System paradigm, 61

THERP, 49
Trust, 89, 90, 92

Wider context, 85, 86
WPAM, 53
Wylfa nuclear power station accident, 11

Zeebrugge car ferry accident, 10